A Mathematician at the BALLPARK

ODDS AND PROBABILITIES FOR BASEBALL FANS

KEN ROSS

PI PRESS
NEW YORK

PI PRESS

An Imprint of Pearson Education, Inc.
1185 Avenue of the Americas, New York, New York 10011

Pi Press offers discounts for bulk purchases. For information
contact U.S. Corporate and Government Sales, 1-800-382-3419,
corpsales@pearsontechgroup.com. For sales outside the U.S.A.,
please contact: International Sales, 1-317-581-3793,
international@pearsontechgroup.com.

Printed in the United States of America

Second Printing

Library of Congress Cataloging-in-Publication Data
A CIP catalog record for this book can be obtained from
the Library of Congress.

Pi Press books are listed at www.pipress.net

ISBN 0-13-147990-3

Pearson Education Ltd.
Pearson Education Australia Pty., Limited
Pearson Education South Asia, Pte. Ltd.
Pearson Education Asia Ltd.
Pearson Education Canada, Ltd.
Pearson Educación de Mexico, S.A. de C.V.
Pearson Education—Japan
Pearson Malaysia SDN BHD

Baseball fans love numbers. They love to swirl them around in their mouths like Bordeaux wine.

PAT CONROY, WRITER

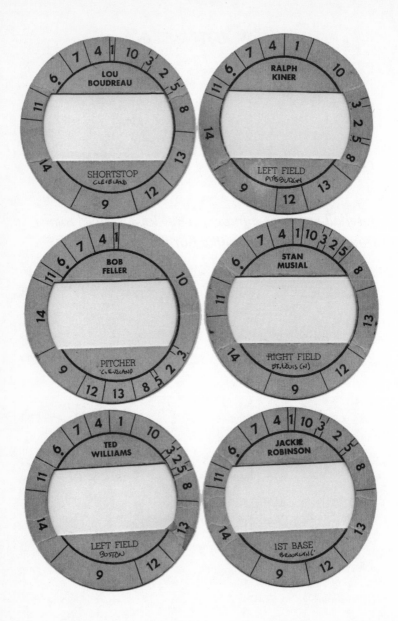

Contents

Preface

When I was six, my father gave me a bright red scorebook that opened my heart to the game of baseball.[1]

DORIS KEARNS GOODWIN, WRITER AND HISTORIAN

My daughters first came to the ballpark when they were so young, they couldn't even track what was going on. Their first understanding of baseball was that innings were amounts of time between dinner (after the 3rd inning) and Fudgesicles (after the 6th). The next insight was that when the ball sailed over the fence, that was a good thing—people stood up and cheered; and the billboard cow, advertising a local dairy, nodded her pressboard head in approval. By the time the girls were nine and six, they were following the plays and learning to keep score.

Some of their questions were naive: "Dad, how come everyone is yelling 'Go Die, Ump?'"

"That's 'Good Eye, Ump,' dear, and we're being sarcastic."

"Dad, what's sarcastic?"

Other questions invited a discussion: "Dad, how come people are yelling 'Yer Due!'?"

"Well, that requires looking at batting averages. That guy's batting average is 290 and that means that, until recently, he's been getting hits about 29 percent of the time. But, in the past week, he's had very few hits, so it seems like he's overdue for some hits. One could question that logic because —"

"Thanks, Dad. Would you like some Cracker Jacks?"

Even in the olden days of my youth, baseball reached all parts of the United States and also other countries, especially Japan and Latin America. However, in America, Major League baseball was regarded as the ultimate level of baseball.

In those ancient days before satellite communications, the Major Leagues were physically restricted to the Northeast and Midwest, a relatively small part of the country. From 1903 to 1952 there were sixteen of these teams located in ten cities in the parallelogram with corners at Chicago, Boston, Washington, DC, and St. Louis.

Many baseball fans and baseball writers from this era grew up in big cities like New York, Chicago, Philadelphia, Detroit, and Boston, so they experienced Major League baseball first-hand. Or at least they were able to follow the regular season by listening to the radio. For example, many baseball fans in the South listened to St. Louis Cardinals' games on the radio. My situation was quite different.

I grew up in Utah, which seemed about as remote from the center of baseball action as a kid could get. I was aware of the brazenly named "World Series." But, for a long time, I didn't even wonder how the teams got to be in the World Series. It seemed like being in New York was a big advantage, if not a requirement, though teams like the Cleveland Indians made cameo appearances. I didn't question this any more than I would have questioned how my parents came to be married or that Roosevelt and Stalin came to be world leaders. It had always been so. Even out in Utah, cool teachers would let their classes listen to the World Series on the radio; my cool teacher was a young lady who taught history. Thanks, Miss Lake.

The real professional baseball team in my home town was our local Salt Lake City Bees, a team in the class C Pioneer League. This league included teams like the Idaho Falls Russets (think potatoes) and the Billings Mustangs (think horses, not cars). I attended many games and I listened to all the other games on my radio, carefully hidden under the bed covers when necessary. I remember recalculating hitters' batting averages—long division with a *pencil!*—and then checking the Salt Lake Tribune the next morning to see if I'd done them right.

Hub Kittle was the pitcher-manager of the Bees. He was a legend in the minor leagues for decades. On August 27, 1980, as a minor league pitching coach, he took the mound for Springfield (Illinois) versus Iowa, teams in the American Association. He was six months past his 63rd birthday and was the oldest player ever in organized baseball. He retired the Iowa batters in the first inning on two flies and a groundout; he threw one pitch in the second inning before leaving the game.

The most interesting ballplayer for the Bees was the catcher, Gus Triandos. He was nicknamed "Tremendous Triandos" because of his rather solid build. He was powerful, but not swift. Indeed, he was the only Major League player to play over 1000 games, mostly with Baltimore, and end up with a perfect stolen-base record, 1 for 1. In addition, he holds the Major League record with 1,206 consecutive games without being caught stealing. In one inconsequential game, near the end of a season, he decided to amble from first base to second base. The catcher was so surprised that he did not throw. (Uncontested stolen bases counted in those days.)

The Salt Lake City Bees, and the World Series via radio, were not enough to satiate my appetite for baseball. My favorite board game was All-Star Baseball,[2] a game designed to simulate real baseball. This was a precomputer nonelectronic game that modeled the hitting of well-known ballplayers of the 1940s, including Cleveland shortstop and manager Lou Boudreau, Yankee center fielder Joe DiMaggio, the Pittsburgh slugger Ralph Kiner, St. Louis Cardinal Stan Musial, Brooklyn infielder Jackie Robinson and the Boston outfielder Ted Williams. Some pitchers were included, like Cleveland pitcher Bob Feller. For each such player, there was a cardboard disk that was to be placed over a spinner; see the samples on page vi. The edge of the disk was marked off with coded possible outcomes like Home Run, Triple, Double, Base on Balls, Strike Out, Ground Out, and Fly Out. There were two versions of the Single, the little single on which any runners advanced one base and the big single on which any runners advanced two bases. For example, the regions 1, 5, 9, and 10 represented Home Runs, Triples, Bases on Balls, and Strike Outs. The regions of these outcomes were proportional to the baseball player's real lifetime performances. Thus sluggers like Ted Williams and Ralph Kiner had relatively large regions for Home Runs, while the pitcher Bob Feller had large regions for Strikeouts and other ways to make outs. All of the players had miniscule regions for Triples.

The idea of the game was to create a couple of teams and then have them play, at-bat by at-bat, until the game was completed. Each hitter's at-bat was determined by flip-

---

[2] © Cadaco-Ellis, Inc. circa 1949.

ping the spinner, using the disk for that player. The bigger the Home Run region on the disk, the more likely it was that the spinner would land on Home Run! Two players could create a team, or one could play both teams by one-self. I learned about, and got a good feel for, probability by playing this game over and over. And over.

The probabilities on the cards were for players' entire careers and did not take into account who the opposing pitchers were, strategies used by both teams, the size and shape of the ballparks, and many other factors that affect hitters' performances. As we shall see in this book, proba-babilistic analyses become more realistic and valuable when more such factors are taken into account. In fact, later versions of the game, All-Star Baseball, took into account some basic strategies of the game as well as the bat-ters' lifetime records.

Thanks to baseball, I became comfortable with, and reasonably proficient at, elementary probability. When I later encountered probability as a formal concept, I found it completely intuitive. This contrasts vividly with how many people feel about probability, so I credit the game for my good intuition. As I think about it, I realize that my first statistical observation was the positive correlation, a word I didn't know then, between home runs and strike-outs of the players; sluggers tend to strike out more.

Like other baseball fans, I've had different kinds of en-counters with baseball: listening to the games on the radio, attending minor league games, later watching games on TV, reading baseball summaries in the newspaper, and of course my fantasies with my board game. In all of these en-counters, I was bombarded with statistics like players' bat-ting averages. Why? Do these statistics tell me exactly what to expect? Is baseball predictable? The short answer is: No, past performance does not predict future results in the short run as any baseball fan and some wise financial investors know. Nevertheless, we appreciate the statistics because they do give us an idea of what to expect. Moreover, in the long run, there are patterns. In fact, there would be patterns even if the players were robots with the same skills but no memory. Probability and statistics are tools that allow us to compare players' performances with how they would do if they were robots. This helps us decide whether their recent changes in performance are due to real changes or just to randomness.

My family and I were big Eugene Emeralds baseball fans in the 1970s. During the 1969-1973 seasons, the Emeralds were the AAA farm team, in the Pacific Coast League, for the Philadelphia Phillies. Near the beginning of the second game of an Emeralds' double-header, a certain batter from the opposing team came to the plate. I stood up, handed my scorebook to my wife, and said, "Here. Hold this. I'm going to get a foul ball." In the next moment, the ball bounced off a wall and into my mitt just as I expected. My family was duly impressed, but I had noticed that during the first game all of this guy's foul balls had been going to the same part of the stands above where we were sitting. So I positioned myself in the appropriate place. That was the only foul ball I ever recovered at a professional ballgame. Even though this event may have given my family the impression that baseball is perfectly predictable, it generally isn't.

One day at the ballpark, we were sitting near a fellow from Philadelphia and his son. He explained that he was on a business trip, but that he had brought his 13-year-old son so that he could see a minor league baseball game. I paled and stammered, "I'm 37 years old and I've never been to a Major League game!" My young daughters promptly planned a trip to Candlestick Park in San Francisco, the nearest Major League park. I've now been to games in about 15 different Major League ballparks. We still follow the Eugene Emeralds, which has been an A-level farm team for several Major League teams. And my daughters have even made me become a Seattle Mariners fan.

I'm a mathematician, and for thirty years I taught probability and elementary statistics at a college. I love probability and baseball. I enjoy casual gambling at casinos, especially blackjack, but I know enough about my long-term expectations, as I'll explain in Chapter 7, to avoid investing much time or money in this. Some examples in the book involve roulette and the lottery, but I take a close look at gambling on baseball in Chapter 4.

Odds are a natural tool for gamblers but are something of a step-child of probability, so I had only casual encounters with odds over the years. In my journey through the garden of probability, with a baseball mitt on my hand, I went from probability to odds, but I'm convinced that most baseball fans are more comfortable with odds than with probability. You'll know intuitively many of the concepts discussed in this book. I will make the concepts more precise and then use them to explain some results in probability that have interesting applications to baseball. Probability is a wonderful window into the workings of baseball, gambling, and, sometimes it seems to me, life itself.

Who's the Best Hitter?
Averages

The box score, being modestly arcane, is a matter of intense in-difference, if not irritation, to the non-fan. To the baseball-bitten, it is not only informative, pictorial, and gossipy but lovely in aesthetic structure.

ROGER ANGELL, AUTHOR

You don't need to be a baseball fan to have heard that a certain player is having a good season and hitting .300. Or perhaps a bad season and not even batting .230. These numbers are a kind of average. In baseball, a player's **batting average**, *AVG*, over a particular period of time is the number of hits, *H*, divided by the number of official at-bats, *AB*, during that period. That is,

$$AVG = \frac{H}{AB}.$$

For example, Babe Ruth's best season, from the point of view of batting average, was in 1923 when he got 205 hits in 522 official at-bats. Thus $H = 205$, $AB = 522$, and

$$AVG = \frac{H}{AB} = \frac{205}{522} \approx .393.$$

I will use $\approx$ whenever an equality is just an approximation. To seven decimal places, this average is .3927203, but it is traditional to give batting averages to three places. So, the average .393 is Babe Ruth's batting average for the 1923 season. Babe Ruth's lifetime batting average isn't so shabby either. It is .342 because, for his lifetime, $AB = 8399$, $H = 2873$, and

$$AVG = \frac{H}{AB} = \frac{2873}{8399} \approx .342.$$

By the way, the phrase "batting average" is an unfortunate choice. It should be "hitting proportion" because it gives the proportion of official at-bats that were hits. Thus throughout his career, about .342 of Babe Ruth's official at-bats were hits. Or, if you prefer percentages, about 34.2 percent of his official at-bats were hits. Of course, "batting average" is absolutely standard and all baseball fans know what it means, so I won't fight it.

There are other numbers that we can think of as averages that many baseball strategists feel more accurately measure hitting effectiveness. In their book [36, *Baseball Dynasties,* page 146], Rob Neyer and Eddie Epstein are emphatic: "Why do people have such a hard time letting go of batting average? On-base percentage and slugging percentage are **more** important, **more** significant, **more** meaningful, **more everything** than batting average. Batting average is a red herring. Let go of it, friends. It's not that important." In case you missed that, they later say, on page 282: "Dear Reader, if you learn nothing else from this book, please learn the **fact** that batting average is not the most important offensive number."

Let's look more closely at some of these other sorts of averages, which came into prominence in the 1980s. **Slugging percentage**, *SLG*, takes into account the **total** number of **bases**, *TB*, that is,

$$TB = 1B + 2 \times 2B + 3 \times 3B + 4 \times HR,$$

where $1B$ is the number of singles, $2B$ is the number of doubles, $3B$ is the number of triples, and HR is the number of home runs. Then the slugging percentage is

$$SLG = \frac{TB}{AB}.$$

Here's another formula for *SLG:*

$$SLG = \frac{H + 2B + 2 \times 3B + 3 \times HR}{AB},$$

which is easier to use when you're not provided the number $1B$ of singles.

In 1923, Babe Ruth had 106 singles, 45 doubles, 13 triples, and 41 home runs. Therefore he had

$$TB = 106 + 2 \times 45 + 3 \times 13 + 4 \times 41 = 399$$

total bases, and his slugging percentage was

$$SLG = \frac{TB}{AB} = \frac{399}{522} \approx .764.$$

This is spectacular, though his slugging percentage was .847 in 1920, which was a record up to the year 2001, when Barry Bonds finally broke it. For Babe Ruth's career, $1B = 1517$, $2B = 506$, $3B = 136$, and $HR = 714$, so

$$TB = 1517 + 2 \times 506 + 3 \times 136 + 4 \times 714 = 5793,$$

and therefore Babe Ruth's lifetime slugging percentage was

$$SLG = \frac{TB}{AB} = \frac{5793}{8399} \approx .690,$$

which is by far the highest in Major League history, assuming at least 3000 plate appearances.

Let's compare the batting average, AVG, and the slugging percentage, SLG. Because the number of hits is the total of the numbers of singles, doubles, triples, and home runs, we can write $H = 1B + 2B + 3B + HR$. Thus we have

$$AVG = \frac{1B + 2B + 3B + HR}{AB},$$

$$SLG = \frac{1B + 2 \times 2B + 3 \times 3B + 4 \times HR}{AB}.$$

The only difference between these formulas is that SLG gives doubles double-the-value, triples triple-the-value, and home runs four times the value.

Recall that AB represents the "official" at-bats. This does not take into account the number of bases on balls, BB, the number of times the batter is hit by a pitched ball, HBP, or the number of times the batter hits a sacrifice fly, SF. It is sometimes convenient to calculate batter effectiveness in terms of *all* plate appearances. The number of all **plate appearances** is

$$PA = AB + BB + HBP + SF.$$

We might regard the added terms, *BB*, *HBP*, and *SF* as the "unofficial" at-bats.

Another popular measure of batter effectiveness is the **on-base percentage,** *OBP*, defined by

$$OBP = \frac{H + BB + HBP}{PA}.$$

This is the proportion of plate appearances where the player gets on base. Note that home runs count as "on base," even though the batter just touches the bases on his way back to home plate.

The numbers *HBP* and *SF* are not readily available for players, like Babe Ruth, who played before statisticians and mathematicians really started analyzing the game. Let's calculate the lifetime on-base percentage, *OBP*, for someone a little closer to our era, Pete Rose. His career numbers are $AB = 14{,}053$, $H = 4256$, $TB = 5752$, $BB = 1566$, $HBP = 107$, and $SF = 79$. So

$$PA = AB + BB + HBP + SF = 14{,}053 + 1566 \\ + 107 + 79 = 15{,}805$$

and Pete Rose's lifetime on-base percentage is

$$\begin{aligned} OBP &= \frac{H + BB + HBP}{PA} \\ &= \frac{4256 + 1566 + 107}{15{,}805} \approx .375. \end{aligned}$$

His number of hits, $H = 4256$, is the all-time record. So is his number of at-bats $AB = 14{,}053$; next is Hank Aaron who had 12,364 lifetime at-bats. Pete's lifetime batting average was

$$AVG = \frac{4256}{14{,}053} \approx .303.$$

This is pretty good, but as Ted Williams put it, "Baseball is the only field of endeavor in which a man can succeed three times out of ten and be considered a good performer."

Each of AVG, SLG, and OBP is an interesting statistic that provides useful information about a batter's effectiveness. But none of these statistics provides a single number that summarizes batters' overall effectiveness. AVG is the oldest established statistic and is invariably the one used to determine "batting champions." But AVG doesn't take into account extra-base hits, much less home runs, runs-batted-in, bases on balls, and other important measures of batter effectiveness. SLG certainly takes into account extra-base hits, but it gives *too much* credit for them. Is a triple *really* three times as valuable as a single? For that matter, is a home run four times as valuable as a single? Getting on base is crucial, so AVG does measure an important statistic. But there are other ways to get on base, most notably bases on balls, so the on-base percentage OBP is also a useful statistic. In other words, AVG, SLG, and OBP are all useful statistics, but none of them is close to reflecting the whole picture.

Another measure of a batter's offensive contribution is **On-Base Plus Slugging** (OPS). The formula is very simple:

$$OPS = SLG + OBP.$$

This is one of the statistics championed by Pete Palmer, one of the two gurus of baseball statistics over the past 20 years; the other one is Bill James. *OPS* is becoming increasingly popular because it's easy to calculate. Moreover, though it may appear as an artificial statistic, it correlates quite well with some lesser known, but more complicated, statistics that are more realistic measures of overall hitting prowess than *AVG, SLG,* or *OBP.*

Let's return to the enigmatic Pete Rose. To calculate his lifetime slugging percentage *SLG,* we need the breakdown of his 14,053 hits: $1B = 3215, 2B = 746, 3B = 72$, and $HR = 223$. Then Pete's

$$TB = 3215 + 2 \times 746 + 3 \times 72 + 4 \times 223 = 5815,$$

so

$$SLG = \frac{TB}{AB} = \frac{5815}{14,053} \approx .414.$$

Now Pete's lifetime

$$OPS = SLG + OBP = .414 + .375 = .789.$$

It's an interesting fact that Pete Rose does not rank in the top 100 of all ballplayers in any of the categories *AVG, SLG, OBP,* or *OPS.*

We now turn our attention to Barry Bonds, regarded by many as the best hitter of our time. In 2001, his statistics were $AB = 476, H = 156, 1B = 49, 2B = 32, 3B = 2, HR = 73, BB = 177, HBP = 9$, and $SF = 2$. Bonds' 73 home runs established a new season record. We have

$$TB = 49 + 2 \times 32 + 3 \times 2 + 4 \times 73 = 411,$$

so

$$AVG = \tfrac{H}{AB} = \tfrac{156}{476} \approx .328,$$

$$SLG = \tfrac{TB}{AB} = \tfrac{411}{476} \approx .863,$$

$$PA = AB + BB + HBP + SF$$
$$= 476 + 177 + 9 + 2 = 664,$$

$$OBP = \tfrac{H + BB + HBP}{PA} = \tfrac{156 + 177 + 9}{664} = \tfrac{342}{664} \approx .515,$$

$$OPS = SLG + OBP \approx .863 + .515 \approx 1.379.$$

In that last line, it looks like I can't add. But, the approximation 1.379 is correct, because $SLG \approx .863445$ and $OBP \approx .515060$, and so $OPS \approx 1.378505$, which rounds up to 1.379. Bonds' slugging percentage, .863, for the season exceeds the older record of .847 set by Babe Ruth in 1920. His OPS of 1.379 just beats Babe Ruth's 1.378 set in 1920. Of course, the Babe never heard of the concept OPS.

In 2002, Barry Bonds' statistics were $AB = 403$, $H = 149$, $1B = 70$, $2B = 31$, $3B = 2$, $HR = 46$, $BB = 198$, $HBP = 9$, and $SF = 2$. This time

$$TB = 70 + 2 \times 31 + 3 \times 2 + 4 \times 46 = 322,$$

so

$$AVG = \tfrac{H}{AB} = \tfrac{149}{403} \approx .370,$$

$$SLG = \tfrac{TB}{AB} = \tfrac{322}{403} \approx .799,$$

$$PA = AB + BB + HBP + SF$$
$$= 403 + 198 + 9 + 2 = 612,$$

$$OBP = \tfrac{H + BB + HBP}{PA} = \tfrac{149 + 198 + 9}{612} = \tfrac{356}{612} \approx .582,$$

$$OPS = SLG + OBP \approx .799 + .582 \approx 1.381.$$

In 2002, Bonds' $AVG = .370$ led the majors. More impressive, his $OBP = .582$ exceeds Ted Williams' .553, a record that goes back to 1941. His *OPS* is also a new record, just beating out his own record of 2001. Bonds is a fantastic player, but it isn't fair to say that he is as great as Babe Ruth because Ruth was the dominant player for nearly 20 years and changed the character of the game.

So much for household names. Let's consider some players who haven't yet ascended to the pantheon. Consider the 2003 regular-season offensive numbers for the two outstanding Japanese outfielders: the Yankee Hideki Matsui and the Mariner Ichiro Suzuki. Some of their numbers are in the next table.

Hideki Matsui and Ichiro Suzuki

	AB	H	2B	3B	HR	BB	HBP	SF
H. Matsui	623	179	42	1	16	63	3	6
Ichiro S.	679	212	29	8	13	36	6	1

For Matsui,

$$AVG = \frac{H}{AB} = \frac{179}{623} \approx .287$$

and

$$SLG = \frac{H + 2B + 2 \times 3B + 3 \times HR}{AB}$$
$$= \frac{179 + 42 + 2 + 48}{623} = \frac{271}{623} \approx .435.$$

Because his number of plate appearances is

$$PA = AB + BB + HBP + SF$$
$$= 623 + 63 + 3 + 6 = 695,$$

Matsui's

$$OBP = \frac{H + BB + HBP}{PA} = \frac{179 + 63 + 3}{695} \approx .353.$$

Finally, Matsui's

$$OPS = SLG + OBP \approx .435 + .353 = .788.$$

We list these numbers in the next table, as well as Matsui's number R of runs scored, number RBI of runs batted in, number SO of strike-outs, number SB of stolen bases, number CS of times caught trying to steal a base, and number E of fielding errors. Exactly the same formulas were used to calculate the numbers for Ichiro, as shown in the following table.

Hideki Matsui and Ichiro Suzuki

	AVG	SLG	OBP	OPS	R	RBI	SO	SB	CS	E
H. Matsui	.287	.435	.353	.788	82	106	86	2	2	8
Ichiro S.	.312	.436	.352	.788	111	62	69	34	8	2

On the basis of the ever-popular *AVG*, Ichiro had a substantially better offensive season than Matsui. But their more significant numbers *SLG*, *OBP*, and *OPS* are amazingly close to each other. These numbers suggest that they are equally effective offensively, though one can always argue over related issues. For example, Matsui had more walks than Ichiro, and Matsui had an outstanding postseason. But the walks were already taken into account in calculating *OBP* and hence *OPS*. Matsui fans might point out that Matsui had significantly more *RBI*s, but that's not surprising since Ichiro was a lead-off batter. One offensive statistic that's been overlooked so far is stolen bases. Ichiro stole 34 in 42 attempts, while Matsui stole two in four attempts, so Ichiro wins out here. Also, Ichiro's fielding was better with two errors compared with Matsui's eight. In fact, Ichiro had only six errors in the three years 2001–2003.

In their important book [42, *The Hidden Game of Baseball*], John Thorn and Pete Palmer prefer the *product* of *OBP* and *SLG,* which is called the **Batter's Run Average** or

$$BRA = OBP \times SLG.$$

They argue that *BRA* is a better measure of offensive power than the *sum OPS = OBP + SLG* because it correlates better with the number of runs produced. I don't disagree, but *OPS* and *BRA* are closely correlated and, moreover, *OPS* is easier to calculate, easier to relate to, and more popular.

Some authors use *SLOB* in place of *BRA.* See, for example, [7]. There's actually some merit to *SLOB.* This is a catchy, if not neat, acronym, and it's easy to remember that

$$SLOB = \mathbf{SLG} \times \mathbf{OBP}.$$

However, it is generally a bad idea to fight established notation, and *BRA* is the standard. By the way, some authors (e.g., [40]) call *SLG* and *OBP* the slugging average and the on-base average, because they are not percentages. I agree, but they aren't averages either.

In much of the nonbaseball world, "average" refers to what you get if you sum a bunch of numbers and divide by the number of such numbers. Thus the average salary of players on a baseball team can be calculated by summing the players' salaries and dividing by the number of players. For example, if there were 26 players and the annual player payroll is \$43,500,000, then the average salary would be

$$\$\frac{43,500,000}{26} \approx \$1,673,000.$$

Figuring out who is the best hitter is not a simple task, even when one understands what the various averages really mean. But some statistical phenomena can make the task seem impossible. We end this chapter with one such phenomenon, called **Simpson's paradox,** which was first observed in the context of economics and that pops up in unexpected places. Let's consider two recent ballplayers, Derek Jeter and David Justice. In 1995,

$$\text{Jeter's } AVG = \frac{H}{AB} = \frac{12}{48} = .250 \quad \text{and}$$
$$\text{Justice's } AVG = \frac{H}{AB} = \frac{104}{411} \approx .253.$$

Justice also had a higher batting average in 1996:

$$\text{Jeter's } AVG = \frac{H}{AB} = \frac{183}{582} \approx .314 \quad \text{and}$$
$$\text{Justice's } AVG = \frac{H}{AB} = \frac{45}{140} \approx .321.$$

It seems reasonable to conclude that, just based on batting averages, Justice did better than Jeter over these two years. However, if we combine the two years, we find that

$$\text{Jeter's } AVG = \frac{12 + 183}{48 + 582} = \frac{195}{630} \approx .310,$$

while

$$\text{Justice's } AVG = \frac{104 + 45}{411 + 140} = \frac{149}{551} \approx .270.$$

This bizarre phenomenon seems to occur for some pair of interesting ballplayers about once a year. To make this particular story even more delicious, observe that Justice beat out Jeter again in 1997:

$$\text{Jeter's } AVG = \frac{190}{654} \approx .291 \quad \text{and}$$

$$\text{Justice's } AVG = \frac{163}{495} \approx .329.$$

But if we combine all three years, we get

$$\text{Jeter's } AVG = \frac{12 + 183 + 190}{48 + 582 + 654} = \frac{385}{1284} \approx .300,$$

while

$$\text{Justice's } AVG = \frac{104 + 45 + 163}{411 + 140 + 495} = \frac{312}{1046} \approx .298.$$

You could say that there's no justice for Justice!

Simpson's paradox isn't a true paradox, but it's called a "paradox" because it's counterintuitive and even distressing. When batting averages are combined, one adds both the numerators and the denominators because that's the way to get the total number of hits and at-bats for the longer period of time. This is different from adding the actual numerical fractions. Since Jeter's 1995 and 1996 averages, .250 and .314, are less than Justice's averages, .253

and .321, Jeter's sum .564 is certainly less than Justice's sum .574. But these sums are meaningless in this context, and one doesn't get the longer-term averages in this way.

Who's the best hitter? One reason that this question causes so many arguments is that there's so many ways to measure hitting excellence. Which is the better measure depends in part on what are, in your judgment, the most important aspects of hitting. None of our averages will give you a simple absolute answer as to who is the best hitter, but they do have the power to reveal the truth about why a hitter is good.

Chapter 2

But Which Team Are You Betting On?
Odds & Probabilities

A million-to-one shot came in. Hell froze over. A month of Sundays hit the calendar. Don Larsen today pitched a no-hit, no-run, no-man-reach-first game in a World Series.

SHIRLEY POVICH, SPORTSWRITER (1956)

Suppose that early some October you and I agree that the New York Yankees are 3 to 2 favorites to win the World Series. We might say that the **odds** are 3:2 that the Yankees will win the series; henceforth I'll use the colon : in this way to signify odds. What do we mean by these odds?

Based on our knowledge of the teams and any hunches, intuition, or superstition that is part of our mental arsenal, we believe that there's about a 3 in 5 chance or likelihood that the Yankees will win the series. Put another way, if the *exact same contest* could somehow be repeated over and over, then we'd expect the ratio of the Yankees wins to their losses to be 3 to 2. For example, if this imaginary experiment could be repeated 100 times, we'd expect the Yankees to win about 60 of the series and lose about 40 of them.

Now suppose you would like to bet that the Yankees will win this series. Because we agree the odds are 3:2 that the Yankees will win, a fair bet would be as follows. If you'll bet or risk $3, and if the Yankees win, I will pay you $2. (If you had given me your $3 for safe keeping, I would return it and pay you the extra $2.) If, on the other hand, the Yankees had lost the series, then I would get (or keep) your $3. This seems *fair* to both of us, because if you could make such a bet five times, you'd expect to win $2 three times and lose $3 two times. We'd end up even.

Similarly, if you chose to bet against the Yankees, then your odds of winning would be 2:3. You would bet or risk $2 and you would get $3 if the Yankees lost. This is again *fair*, because if you could repeat this five times, you'd expect to win $3 twice and lose $2 three times. I will return to the idea of fair bets in Chapter 3.

We need to be able to talk about, and work with, odds in general. So I need to introduce some notation. Since our most interesting applications will involve betting, I'll use the notation $b{:}a$. Here b is to remind us of the original *bet*. If you'd like, you can view a as a reminder of what you hope to *acquire* if you win. In our example with the World Series, $b = 3$ and $a = 2$ if we bet for the Yankees, and $b = 2$ and $a = 3$ if we bet against them. In all cases, odds $b{:}a$ in favor of an outcome indicate that we believe that the chances or likelihood of the outcome is $\frac{b}{b+a}$. For example, if the odds are 3:2, then the chances are $\frac{3}{3+2} = \frac{3}{5}$, just as we agreed earlier.

The case where we don't agree on the odds is a little more interesting.

Suppose, for example, you feel the New York Yankees are favored 3:2 to win the World Series, and I feel they are favored 8:5. Now there is no way you can bet on the Yankees and have us both feel it's *fair*. The odds 8:5 are better (from the Yankees' point of view) than the odds 3:2, because the ratio $\frac{8}{5} = 1.6$ is bigger than the ratio $\frac{3}{2} = 1.5$.

Suppose you bet on the Yankees based on the odds 3:2. Because I believe that the true odds are 8:5, over thirteen games I'd expect you to win $2 eight times and lose $3 five times, for a net gain of $1. In other words, over the long haul, I would feel that you had an unfair advantage.

On the other hand, suppose you bet on the Yankees based on the odds 8:5. Because you believe the true odds are 3:2, over five games you'd expect to win $5 three times and lose $8 two times, for a net loss of $1. Thus you would feel that you were at an unfair disadvantage. To see this another way, note that the odds 3:2 are the same as the odds 7.5:5, so if you were to bet $8 on the Yankees based on the odds 8:5, you would be betting or risking 50 cents more than you should have to.

If we each really believe in our different odds, and we want to remain friends, we should agree not to bet.

If you gamble at casinos, online, or with other professionals who intend to make money, they will not simultaneously offer odds *b:a* in favor of an outcome and the reversed odds *a:b* against the outcome. If they did so because they felt that these were the correct odds, then they would be in a *fair* game whether the gambler bet in favor or against the outcome. In that case, they wouldn't make any money. They adjust the odds that they offer so that they have an advantage whether or not the outcome occurs. We will analyze this situation in some detail in Chapter 4, in the setting of betting on baseball games.

The advantage of using odds is that they encode the betting strategy directly. Moreover, their meaning is pretty straightforward. The disadvantage is that there are many questions that are difficult, if not impossible, to answer using the language of odds. This is why mathematicians and statisticians almost always work with probability instead.

Suppose that one day, during the regular season, we are given that the Atlanta Braves are favored 3:2 to win their game and the Chicago Cubs are favored 4:3 to win their game. The Braves and Cubs are not playing each other. You want to bet that at least one of the teams will win their game. On the other hand, I want to bet that *both* teams win their games. What are the appropriate odds for your bet and my bet? The answers are 29:6 and 12:23. That is, the odds that at least one of the teams wins are 29:6, while the odds are 23:12 *against* both teams winning. Where the heck did I get these odds from? It was easy, but I had to restate and solve the problems in probability, and then translate the answers back into odds. To figure these out just using odds would be like trying to add or multiply numbers using Roman numerals.

Probability is the wonderful part of mathematics that helps us understand statistics, which is the tool to help understand the past and predict the future.

When the odds are $b:a$ for an outcome, then the "chance" or "likelihood" of that outcome is $\frac{b}{b+a}$. The mathematical terminology for "chance" or "likelihood" is **probability,** which we'll often abbreviate as **Pr.** For example, if the Yankees are 3:2 favorites to win the World Series, we could write

$$\mathbf{Pr}(\text{Yankees win the series}) = \frac{3}{5} = 0.60,$$

$$\mathbf{Pr}(\text{Yankees lose the series}) = \frac{2}{5} = 0.40.$$

As in the discussion about the Braves and Cubs, in probability we are often interested in several possible outcomes rather than just one. Also, we are interested in obtaining probabilities of some outcomes in terms of probabilities of other outcomes. To facilitate this new way of thinking, we really need some efficient notation for outcomes of interest. In probability theory, these outcomes are called **events** and are usually denoted by capital letters like E and F, unless giving them different names will help us remember what they stand for.

At the risk of flogging a dead horse, let's consider again the World Series where the Yankees are 3:2 favorites to win the series. We might abbreviate

E = event that the Yankees win the series,

F = event that the Yankees lose the series.

Then the probability that the Yankees will win the series is $\mathbf{Pr}(E) = 0.60$, and the probability that they will lose is $\mathbf{Pr}(F) = 0.40$. Note that the sum of these two probabilities is 1.00, and *exactly one* of these events is sure to occur. In words, the Yankees will either win the series or they will lose the series, but not both. This observation illustrates one of the key properties of probabilities.

Consider a World Series again, in which we don't care whether the Yankees are involved. Instead, we focus on several possible outcomes of interest, namely the number of games the series will take. There are four possible outcomes, depending on whether the series lasts four, five, six, or seven games. Ordinarily, I would name these events E_1, E_2, E_3, and E_4 because there are four of them; the little subscripts 1, 2, 3, and 4 are there to distinguish the events. However, I prefer the names E_4, E_5, E_6, and E_7 so that the subscripts will remind us how long the series lasts. To be precise,

$$E_4 = \text{event that the series lasts 4 games,}$$

$$E_5 = \text{event that the series lasts 5 games,}$$

and so forth.

We don't know enough about the series to determine the probabilities of these events. We would need to know the teams, starting pitchers, and other information. However, in Chapter 6, we'll give some answers that make sense under certain assumptions that may be more-or-less reasonable. One thing is for sure, though: No matter how we assign the probabilities, and for whatever reasons, we won't believe them unless

$$\Pr(E_4) + \Pr(E_5) + \Pr(E_6) + \Pr(E_7) = 1.00.$$

This has to be true because *exactly one* of these four events is sure to occur: Either the series will last four games, or it will last five games, and so forth. That is, either E_4 will occur, or E_5 will occur, and so forth.

We can make some other evident claims. For example,

$$\Pr(E_4 \text{ or } E_5) = \Pr(E_4) + \Pr(E_5).$$

That is, the probability that the series will last four or five games is the sum of the probability that it will last four games plus the probability that it will last five games. Note that this is clear to us, even though we don't know the actual values of the probabilities under consideration. The formula is true simply because outcomes E_4 and E_5 cannot both occur. When the two events overlap, the appropriate formula is a little more complicated. To illustrate this, consider the events

$$E = \text{series lasts 4 or 5 games,}$$

$$F = \text{series lasts 5 or 6 games.}$$

Then

$$\mathbf{Pr}(E) = \mathbf{Pr}(E_4 \text{ or } E_5) = \mathbf{Pr}(E_4) + \mathbf{Pr}(E_5),$$

and similarly

$$\mathbf{Pr}(F) = \mathbf{Pr}(E_5) + \mathbf{Pr}(E_6).$$

Also, we have

$$\mathbf{Pr}(\text{series lasts 4, 5, or 6 games})$$
$$= \mathbf{Pr}(E \text{ or } F) = \mathbf{Pr}(E_4) + \mathbf{Pr}(E_5) + \mathbf{Pr}(E_6).$$

Thus $\mathbf{Pr}(E \text{ or } F)$ is *not* equal to $\mathbf{Pr}(E) + \mathbf{Pr}(F)$. The trouble is that the sum

$$\mathbf{Pr}(E) + \mathbf{Pr}(F) = [\mathbf{Pr}(E_4) + \mathbf{Pr}(E_5] + [\mathbf{Pr}(E_5) + \mathbf{Pr}(E_6)]$$

counts $\mathbf{Pr}(E_5)$ twice. Because the event E_5 is exactly the event "E and F", we could write

$$\mathbf{Pr}(E \text{ or } F) = \mathbf{Pr}(E) + \mathbf{Pr}(F) - \mathbf{Pr}(E \text{ and } F).$$

It turns out that this formula works in general: To get $\mathbf{Pr}(E \text{ or } F)$, you can add the two probabilities provided you subtract the probability that both events occur. Adding probabilities isn't the same as adding ordinary numbers. This reminds me of Yogi Berra's observation that "baseball's 90 percent mental. The other half is physical."

We're ready to formalize what a **probability** is. It is an assignment of values $\Pr(E)$ to events E with the following basic properties:

1. Each number $\Pr(E)$ is between 0 and 1.00.
2. If we have events E_1, E_2, . . . so that exactly one of them is sure to occur *and* there are no overlaps, then the sum of their probabilities is 1.00. That is,

$$\Pr(E_1) + \Pr(E_2) + \cdots = 1.00.$$

The three dots are the mathematicians' *et cetera*. When we want a name for the number of events, but we don't want to specify a value, we'll use a lower-case letter like n or k. Then we might write Property 2 as follows:

2. If we have events E_1, E_2, . . ., E_n so that exactly one of them is sure to occur *and* there are no overlaps, then the sum of their probabilities is 1.00. That is,

$$\Pr(E_1) + \Pr(E_2) + \cdots + \Pr(E_n) = 1.00.$$

Amazingly, a huge amount of probability theory can be established based only on requirements 1 and 2. Here's a nearly obvious special case of Property 2:

3. For any event E, we have
   $\Pr(E) + \Pr(\text{not } E) = 1.00.$

This is true because exactly one of the events E or "not E" must occur. I've been quietly using this property all along:

either the Yankees will win the World Series or not, and the sum of the probabilities of these two events is 1.00.
Here's a variation on Property 2:

4. If we have events $E_1, E_2, \ldots, E_n$, so that no two of them can occur (no overlaps), then

$$\mathbf{Pr}(E_1 \text{ or } E_2 \text{ or} \cdots \text{or } E_n) = \mathbf{Pr}(E_1) \\ + \mathbf{Pr}(E_2) + \cdots + \mathbf{Pr}(E_n).$$

In particular, $\mathbf{Pr}(E \text{ or } F) = \mathbf{Pr}(E) + \mathbf{Pr}(F)$ if E and F cannot both occur at the same time. We saw this property used on page 20 where, for example, we observed that $\mathbf{Pr}(E_4 \text{ or } E_5) = \mathbf{Pr}(E_4) + \mathbf{Pr}(E_5)$. Here's another property that we illustrated:

5. For any two events E and F,

$$\mathbf{Pr}(E \text{ or } F) = \mathbf{Pr}(E) + \mathbf{Pr}(F) - \mathbf{Pr}(E \text{ and } F).$$

If E and F cannot both occur, then $\mathbf{Pr}(E \text{ and } F) = 0$; so $\mathbf{Pr}(E \text{ or } F) = \mathbf{Pr}(E) + \mathbf{Pr}(F)$, and we are back to Property 4 for two events. I will verify Property 5 at the end of the chapter, though the ideas are implicit in the solution to the question regarding the number of games in a World Series.

We already know how to get a probability from odds.

Odds to Probability If the odds are $b{:}a$ for some outcome, then $\mathbf{Pr}$ (the outcome) $= \frac{b}{b + a}$.

Before we illustrate Property 5 with the Braves and Cubs, I will show how to go from probability back to odds. If the probability that the St. Louis Cardinals will win their division is $\frac{5}{9}$, then the probability that they won't win the division is $\frac{4}{9}$. If this situation was repeatable over and over, we'd expect the Cardinals to win about 5 out of every 9 division titles. That is, the Cardinals are 5:4 favorites to win their division. If we write $\frac{5}{9}$ as $\frac{m}{n}$, then $4 = 9 - 5 = n - m$ and the Cards are $m:(n - m)$ favorites. This thinking works in general, and we have the following rule.

Probability to Odds Suppose that you are given the probability **Pr**(an outcome) as a fraction $\frac{m}{n}$. Then the odds for the outcome are $m:(n - m)$.

Note that $\frac{m}{n}$ is less than 1, so m is less than n, and $n - m$ is bigger than 0. So writing the odds $m:(n - m)$ at least makes sense. These are the right odds, since given these odds, the probability of the outcome is

$$\frac{m}{m + (n - m)} = \frac{m}{n}.$$

(This is the same as $\frac{b}{b + a}$ where $b = m$ and $a = n - m$.)

Let's return to the situation on page 18, where the Atlanta Braves and Chicago Cubs were favored to win their games 3:2 and 4:3, respectively, and we wanted the odds that at least one of the teams would win and also the odds that both would win. If we write

$$B = \text{event that the Braves win,}$$

$$C = \text{event that the Cubs win,}$$

then (using Property 5)

$$\mathbf{Pr}(B \text{ or } C) = \mathbf{Pr}(B) + \mathbf{Pr}(C) - \mathbf{Pr}(B \text{ and } C).$$

From the given odds, we know $\mathbf{Pr}(B) = \frac{3}{5}$ and $\mathbf{Pr}(C) = \frac{4}{7}$. If we knew either $\mathbf{Pr}(B \text{ or } C)$ or $\mathbf{Pr}(B \text{ and } C)$, we'd be in business because we'd be able to calculate the other one using the above formula, and then get the odds using the Probability to Odds rule.

There is good reason to believe *in this case*

$$\mathbf{Pr}(B \text{ and } C) = \mathbf{Pr}(B) \times \mathbf{Pr}(C),$$

because there is no reason to believe that the outcome of the Braves' game will affect the outcome of the Cubs' game. If you will please accept this for now, then the probability that both teams will win is

$$\mathbf{Pr}(B \text{ and } C) = \frac{3}{5} \times \frac{4}{7} = \frac{12}{35}.$$

Translating this back to odds, we see that the odds that both the Braves and the Cubs will win is $12:(35 - 12)$ or 12:23 (think $m = 12$ and $n = 35$), just as I boldly asserted on page 18. Moreover, the probability that at least one team wins is

$$\mathbf{Pr}(B \text{ or } C) = \mathbf{Pr}(B) + \mathbf{Pr}(C) - \mathbf{Pr}(B \text{ and } C)$$
$$= \frac{3}{5} + \frac{4}{7} - \frac{12}{35} = \frac{21}{35} + \frac{20}{35} - \frac{12}{35} = \frac{29}{35},$$

so the odds that the Braves *or* the Cubs win are $29:(35 - 29)$ or 29:6, again as claimed on page 18.

Finally, recall that on page 17, I cheerfully noted that the odds 8:5 were better than the odds 3:2 because the ratio $\frac{8}{5}$ was bigger than the ratio $\frac{3}{2}$. This is perhaps clear to you, but it's even clearer if you observe that

$$\text{with odds } 8\!:\!5, \quad \mathbf{Pr}(\text{outcome}) = \frac{8}{13} \approx 0.615,$$

while

$$\text{with odds } 3\!:\!2, \quad \mathbf{Pr}(\text{outcome}) = \frac{3}{5} = 0.600.$$

In other words, the outcome is more likely with odds 8:5 than with odds 3:2. In general, the following statements are equivalent and both signify that the odds $b{:}a$ are better than the odds $d{:}c$.

$$\text{The ratio } \frac{b}{a} \text{ is bigger than the ratio } \frac{d}{c};$$

the probability $\dfrac{b}{b+a}$ is bigger than the probability $\dfrac{d}{d+c}$.

If you are still dubious, here is an algebraic argument that consists of showing that the following statements are equivalent:

1. $\dfrac{b}{a} > \dfrac{d}{c}$

2. $\dfrac{a}{b} < \dfrac{c}{d}$ (taking reciprocals reverses the inequality sign)

3. $1 + \dfrac{a}{b} < 1 + \dfrac{c}{d}$ (adding 1 to both sides preserves the sign)

4. $\dfrac{b}{b} + \dfrac{a}{b} < \dfrac{d}{d} + \dfrac{c}{d}$ (because $1 = \dfrac{b}{b}$ and $1 = \dfrac{d}{d}$)

5. $\dfrac{b + a}{b} < \dfrac{d + c}{d}$ (algebra)

6. $\dfrac{b}{b + a} > \dfrac{d}{d + c}$ (taking reciprocals reverses the inequality sign)

Some Mathematics

The verification of Property 5 on page 23 involves a little algebra. The events 'E and F' and 'E but not F' don't overlap, and the event E is exactly the same as the following event in double-quotes:

"'E and F' or 'E but not F'".

Therefore, because of Property 4 on page 23, we have

$$\mathbf{Pr}(E) = \mathbf{Pr}(E \text{ and } F) + \mathbf{Pr}(E \text{ but not } F).$$

Similarly, F is exactly the event

"'E and F' or 'F but not E'",

and both of these single-quoted events don't overlap. Again by Property 4 on page 23

$$\mathbf{Pr}(F) = \mathbf{Pr}(E \text{ and } F) + \mathbf{Pr}(F \text{ but not } E).$$

Also, the event 'E or F' is the same as

"'E and F' or 'E but not F' or 'F but not E'",

and no two of these three single-quoted events can occur simultaneously. Therefore, yet again by Property 4 on page 23,

$$\text{Pr}(E \text{ or } F) = \text{Pr}(E \text{ and } F) + \text{Pr}(E \text{ but not } F)$$
$$+ \text{Pr}(F \text{ but not } E).$$

Finally, using a little algebra, we have

$$\text{Pr}(E \text{ or } F) = [\text{Pr}(E \text{ and } F) + \text{Pr}(E \text{ but not } F)]$$
$$+ [\text{Pr}(E \text{ and } F) + \text{Pr}(F \text{ but not } E)] - \text{Pr}(E \text{ and } F)$$
$$= \text{Pr}(E) + \text{Pr}(F) - \text{Pr}(E \text{ and } F).$$

By its very nature, each time the probability of some event is given, it carries uncertainty with it. However, the subject probability itself is not a vague guessing game, but a vigorous science as reliable as $2 + 2 = 4$.

Chapter
3

Will You Win
the Lottery?
Expectations

Never win 20 games in a year, because they'll expect you to do it every year.

BILLY LOES, MAJOR LEAGUE PITCHER, 1950–1961

This chapter is concerned with "fair bets" and "fair games," and these ideas are based on the idea of expectation.

I discussed odds and fair bets in Chapter 2. Suppose again that the odds are 3:2 in favor of an outcome that you want to bet on, such as the Yankees winning the World Series. I said that if you'd risk $3 to win $2, then this would be a fair bet because, "if you could make such a bet five times, you'd *expect* to win $2 three times and lose $3 two times." The truth is that this argument is a bit careless. Even if you could repeat the experiment five times, you wouldn't *expect* to win exactly three times. You might be lucky and win four or five times, or you might be unlucky and win less than three times. What I meant was that this was the average situation, so that in the long run you'd be about even.

What we really want to calculate, then, is the long-term *average* expected gain *per bet*. Or loss per bet, but henceforth we'll talk about negative gains instead of losses. This theoretical average is determined by the probabilities of the different possible outcomes. Over the long term, you'd win $2 about $\frac{3}{5}$ of the time and you'd lose (i.e., win –$3) about $\frac{2}{5}$ of the time. Thus the average expected gain *per bet* is

$$\frac{3}{5} \times \$2 + \frac{2}{5} \times (-\$3) = 0.$$

Such an average expected gain is called the **expectation.** When the expectation is 0, as in this case, the game is said to be **fair.** When expectation involves money, as in this case, it is often referred to as the **value** of the game. Speaking of value, George Steinbrenner once explained that "you measure the value of a ballplayer by how many fannies he puts in the seats."

Let's analyze again the example on page 17, where you feel the Yankees are favored 3:2 and I feel they are favored 8:5. If we agree to bet 3:2, then your expectation is 0, as just calculated above; so you would regard the bet as fair. However, my expectation is

$$\frac{8}{13} \times (-\$2) + \frac{5}{13} \times \$3 = -\$\frac{1}{13} \approx -\$0.0769$$
$$\approx -7.69 \text{ cents.}$$

In other words, if we could repeat this experiment over and over, and if I am correct that the true odds are 8:5, then I'd expect an average loss of almost 8 cents per bet. This is why I would feel you had an unfair advantage.

On the other hand, if we agreed to bet 8:5 and if you are correct that the true odds are 3:2, then your expectation would be

$$\frac{3}{5} \times \$5 + \frac{2}{5} \times (-\$8) = -\$\frac{1}{5} = -\$0.20 = -20 \text{ cents.}$$

If we could repeat this experiment over and over, you'd expect to lose an average of about 20 cents per bet. Clearly, you'd feel that this was unfair.

Even in the simple American game of roulette, calculations of expectation get interesting. At each play there are 38 equally likely possible values; two of them are numbered 0 and 00, and the other 36 are numbered 1 through 36. Half of the 36 numbers are red and half of them are black; the special numbers 0 and 00 are colored green. The bettor has many betting options, but except for a couple of special bets involving 0 and 00, the expectations are all the same, namely about −$0.053 or −5.3 cents for every dollar played. I will illustrate this with two different sorts of bets.

You can choose to bet on red, in which case the house gives you 1:1 odds. I use "house" as an abbreviation for the state government or casino that runs the game. You win $1 if the number comes up red and you lose $1 otherwise. Since 18 of the 38 numbers are red, you would win about $\frac{18}{38}$ of the time and lose about $\frac{20}{38}$ of the time. So, your expectation is

$$\frac{18}{38} \times \$1 + \frac{20}{38} \times (-\$1) = -\$\frac{2}{38} \approx -\$0.053$$
$$= -5.3 \text{ cents.}$$

Note that the green numbers 0 and 00 give the house its edge. Without them, the game would be fair because the expectation would be

$$\frac{18}{36} \times \$1 + \frac{18}{36} \times (-\$1) = \$0.00.$$

If the house wanted the current game with 38 outcomes to be fair, it would give you 9:10 odds, so that if you bet $9, you would get $10 if you won. Now your expectation would be

$$\frac{18}{38} \times \$10 + \frac{20}{38} \times (-\$9) = \$0.00.$$

Another option in the real game is to bet on certain sets of four numbers from among 1 through 36. If you do so, the house gives you 1:8 odds. That is, you risk $1 and you gain $8 if you win. This time your expectation is

$$\frac{4}{38} \times \$8 + \frac{34}{38} \times (-\$1) = -\$\frac{2}{38} = -5.3 \text{ cents},$$

which is the same expectation we obtained earlier. Again, the numbers 0 and 00 give the house its edge. Without them, your expectation would be

$$\frac{4}{36} \times \$8 + \frac{32}{36} \times (-\$1) = \$0.00,$$

and the game would be fair.

In these two examples, the situation has been simple, because we only had two possible outcomes. If the true odds in favor of the outcome you are betting on are $b{:}a$, then your **expectation** is

$$\frac{b}{b + a} \times (\text{your gain if you win})$$

$$+ \frac{a}{b + a} \times (\text{your gain if you lose}).$$

Of course, if you lose, your "gain" will be negative. Often, as in our Yankees example, you won't *know* the true odds.

But if you *believe* that *b:a* are the true odds, then you *believe* that your expectation is the sum just given.

In general, there may be different payoffs or gains for several different outcomes. This occurs with most lotteries. As we will see, with a lottery, usually the expectation for every dollar "invested" is poor, but many people play the lottery anyway, because the prospect of winning big is irresistible. In contrast, expectations for some casino games like roulette, although negative, are much more moderate. As another example, a reasonably careful blackjack player can keep his losses down to about 2 cents per dollar invested. On the other hand, the expectation for keno is about –26 cents for every dollar invested. Bettor beware!

In most casino games, the odds favor the house and there's little that players can do but hope that Lady Luck is with them. Exceptions are blackjack and poker, where good knowledgable players can consistently make money.

The situation for gambling on sports is different and more complicated. A bettor who is very knowledgable about a sport should be able to make money, even against professional gamblers. His knowledge should be able to overcome the gambler's built-in edge, called vigorish. The next chapter is devoted to the situation involving bets on individual baseball games.

Expectation makes sense whenever we have numerical values of interest that depend on various outcomes. The numerical values don't need to represent gain or payoff as in our preceding examples. Before I give an example, let me make it clear what I mean by expectation in this more general setting.

Let's start with our familiar setting of two possible gains, one usually being negative, but use a little different notation. Let

E_1 = the event that the desired outcome occurs,

E_2 = the event that the desired outcome does not occur.

If the odds for E_1 are $b{:}a$, then we have

$$\mathbf{Pr}(E_1) = \frac{b}{b + a} \quad \text{and} \quad \mathbf{Pr}(E_2) = \frac{a}{b + a}.$$

Then the equation defining **expectation** on page 32 can be written as

$$\mathbf{Pr}(E_1) \times (\text{value depending on } E_1)$$
$$+ \mathbf{Pr}(E_2) \times (\text{value depending on } E_2).$$

I've struggled to rewrite that nice equation in this peculiar way to help lead us to the general definition of expectation.

Suppose we have a list of the possible outcomes of interest as the events $E_1, E_2, \ldots$, where one of the events is sure to occur and no two can occur. Then the **expectation** is the sum of all the products

$$\mathbf{Pr}(E_1) \times (\text{value depending on } E_1 \text{ occurs}),$$

$$\mathbf{Pr}(E_2) \times (\text{value depending on } E_2 \text{ occurs}),$$

and so forth. If there is a total of three such events $E_1, E_2,$ and E_3, then we add the two products just listed to

$$\mathbf{Pr}(E_3) \times (\text{value depending on } E_3 \text{ occurs}).$$

I'm sure you'd appreciate an example.

Let's return to the example on pages 19–20 concerning the length of the World Series. The events of interest are $E_4, E_5, E_6,$ and E_7, where

E_4 = event that the series lasts 4 games,

E_5 = event that the series lasts 5 games,

and so forth. The values of interest are the lengths of the series, i.e., the subscripts 4, 5, 6, and 7. Let's assume that the two teams are equally matched in every game. This isn't an accurate assumption, but it isn't a crazy one either. If we make this assumption, then it turns out that

$$\Pr(E_4) = 0.125, \quad \Pr(E_5) = 0.250,$$
$$\text{and} \quad \Pr(E_6) = \Pr(E_7) = 0.3125.$$

Note that these four numbers add to 1, as they should. The expectation, or expected number, of games that the series will last is

$$\Pr(E_4) \times 4 + \Pr(E_5) \times 5 + \Pr(E_6) \times 6 + \Pr(E_7) \times 7.$$

As in the general case, this is the sum of the products of the probabilities of the events and the corresponding values of interest. The sum is

$$0.125 \times 4 + 0.250 \times 5$$
$$+ \ 0.3125 \times 6 + 0.3125 \times 7 = 5.8125.$$

If the assumptions are correct, then in the long run the average length of a World Series would be 5.8125, which is close to 6.

It is interesting to compare this with what has happened over the past century. Since 1905 there have been 94 World Series in which the winner was the first team to win 4 games. I've excluded the series from 1919–1921, because the winner had to win 5 games in those series. Seventeen of the 94 series lasted 4 games, 21 lasted 5 games, 21 lasted 6 games, and 35 last 7 games. If we want the average length of a series, we'll need to sum the 17 appearances of 4 (which is 17×4), sum the 21 appearances of 5 (which is 21×5), and so on. Then we divide this sum by 94. The result is

$$\frac{1}{94} \times (17 \times 4 + 21 \times 5 + 21 \times 6 + 35 \times 7)$$

$$= \frac{544}{94} \approx 5.7872.$$

This is remarkably close to the theoretical expected average of 5.8125. It is so close, we might conclude that our theoretical model, where we assumed that the teams were equally likely to win each game, is a good model.

In fact, it is not such a good model if we examine the situation more closely. One should be careful about jumping to conclusions with limited data! Using the probabilities given for E_4 through E_7, we see that the model predicts that about 0.125 of the series will last 4 games, about 0.250 of them will last 5 games, etc. Thus, out of 94 series, the model predicts that about $0.125 \times 94 \approx 12$ of them will last four games, about $0.250 \times 94 \approx 24$ of them will last five games, etc. The results are summarized in the following table.

Lengths of World Series

Length of series →	4	5	6	7
Predicted number	12	24	29	29
Actual number	17	21	21	35

Based on the 94 World Series under discussion, the predicted numbers of series of each length don't agree very well with the actual numbers. In the first place, there are more series of length four than the model predicted. This is undoubtedly true because, in several of the series, one team was substantially stronger than the other team. The differences in the numbers of series of length six and seven are more mysterious. My theory is that the answer is primarily psychological. In the first place, since the series ran at least six games, one team won three of the first five games and the other team won two of them. So the teams were fairly evenly matched. After the five games, the team with the three wins knew it had two chances to wrap up the World Series, while the other team knew it *must* win the sixth game to have a chance at winning at all. For the sixth game, the pressure was more on the team behind in the series, so they probably played with more determination. Also, the manager of the team that is behind in the series might go with the best pitcher available in the sixth game, since there might not be a seventh game. If you have better explanations for the unusually large number of seven-game series, let me know.

In this book, I am using the terms "expectation" and "expected value" in the probabilistic sense. In particular, these are numbers. This is different from the way we use these words in the real world. When you say that you expect a particular World Series to last seven games, you probably don't have a number in mind. You just think the event is likely. When I say that the expected number of games of a World Series is about 5.79, I really mean that the average number of games over a long long time will be close to this number. So the meanings are different, but there's a loose connection between the normal use and the technical use of these words. This is why probabilists use this language.

The lottery is an area where lots of people have high expectations, but probabilists don't. When dollar signs appear in the world of probability, many people get confused quickly. Why else would so many people expect to win the lottery?

I feel like I just struck out Joe DiMaggio with the bases loaded and two outs in the ninth.

MICKEY MCDERMOTT ON HIS WIFE WINNING
$6 MILLION IN THE ARIZONA LOTTERY

Powerball is a popular lottery in many states. A typical arrangement is that five white balls are drawn from among fifty-three balls, numbered 1 through 53, and that one red powerball (PB) is drawn from among forty-two balls, numbered 1 through 42. For example, one of the winning combinations in February 2003 was 1-26-29-38-42 plus the powerball 30. Players select the five numbers plus the powerball number. If their selection matches the day's winning combination, at least in part, then they win a prize. From the player's point of view, there are ten events of interest, nine winning events and the one way to lose. One of the ten events always occurs, and no two of the events can occur. The question is, what is the player's expectation?

The information needed to answer this question is in Column 5 of the next table, which is based on a single bet of $1. For now you may ignore the numbers and strange notation in Columns 2, 3, 4, and 6.

If you only care about each player's expectation, this is the sum of the products, **Pr** × Gain, in column 5, namely

$$\frac{J}{120.5} - 0.803.$$

This sum depends on the jackpot, which will be J million dollars for some J, a number that is always at least 10. When the jackpot is 10 million, then $J = 10$ and your expectation is

Powerball

Event	Number of Possibilities	Probability of the Event	Gain (Prize)	Pr × Gain	Odds of the Event
All 6 numbers	1	0.000000008	J million	$\approx \frac{J}{120.5}$	1:120,526,769
5 white balls	1×41	0.0000003	$100,000	$0.0300	1:2,939,676
4 white balls & PB	$_5C_4 \times _{48}C_1 \times 1 = 240$	0.000002	$5,000	$0.0100	1:502,194
4 white balls	$_5C_4 \times _{48}C_1 \times 41 = 9840$	0.000082	$100	$0.0082	1:12,248
3 white balls & PB	$_5C_3 \times _{48}C_2 \times 1 = 11,280$	0.000094	$100	$0.0094	1:10,684
3 white balls	$_5C_3 \times _{48}C_2 \times 41 = 462,480$	0.003837	$7	$0.0269	1:260
2 white balls & PB	$_5C_2 \times _{48}C_3 \times 1 = 172,960$	0.001435	$7	$0.0100	1:696
1 white ball & PB	$_5C_1 \times _{48}C_4 \times 1 = 972,900$	0.00807	$4	$0.0323	1:123
0 white balls & PB	$_5C_0 \times _{48}C_5 \times 1 = 1,712,304$	0.01421	$3	$0.0426	1:69
None of the above	117,184,724	0.9723	-$1	-$0.9723	35:1
TOTALS	120,526,770	1.0000		$\frac{J}{120.5} - .803$	

$$\approx \$ \frac{10}{120.5} - 0.803 \approx -\$0.72,$$

or about −72 cents. What does this mean? It means that if you could play the same lottery billions of times, always with a 10 million dollar jackpot, your average loss would be about 72 cents for every dollar played. This is a poor expectation, but rather meaningless, because no one can repeat such games billions of times. However, the house's expectation of gaining 72 cents for every dollar played is very meaningful. The house can expect to do very well. Even if someone wins the 10 million dollar jackpot, the house can expect to keep 72 percent of what's played.

If the jackpot is 97 million dollars, then $J = 97$ and the expectation is

$$\approx \frac{97}{120.5} - 0.803 \approx \$0.00.$$

The largest jackpot that I'm aware of was 280 million dollars. At that time, a player's expectation was

$$\approx \frac{280}{120.5} - 0.803 \approx \$1.52.$$

How can the house afford to have the jackpot so big that its expectation is to lose over $1.50 for every dollar played? The answer is that the jackpot didn't get that big unless it was preceded by several days with no jackpots. Even though J figures in the calculations of the expectations on those days, once the house realizes that no one won the jackpot, they can ignore J's contribution to the expectation and rest comfortable that their real expectation on such days will be close to −80 cents for every dollar played. So for those days before the big jackpot was won, the house took in about 80 percent (from 0.803 in the formula $\frac{1}{120.5} - 0.803$) of what was played. Of course, this is an estimate for each day, but it won't be far off. Just as in gambling casinos in Atlantic City and Nevada, a so-called "law of large numbers" protects the house from severe deviations from the expected. Each player risks his "investment," but the house takes essentially no risks.

A few years ago, I went to a regional mathematics meeting. When I arrived, I noticed many local folks congratulating a good friend of mine. So, of course, I asked him what this was all about. That term he was teaching elementary probability at a local college. At the end of class on a particular Wednesday, he mentioned that "if ever you're going to bet on the lottery, now is the time because right now the jackpot is high and your expectation is positive, namely $0.14 for each $1 bet. When he was driving home after work, he was stopped at a stop light right by his favorite market. Even though he doesn't normally play the lottery, he thought to himself, "I should put my money where my mouth is." So he turned around, went back to the market, and bought ten $1 tickets. When he got home, he proudly showed the tickets to his wife and said, tongue-in-cheek, "Look honey. Each of these tickets is worth $1.14 and I only paid a dollar each for them."

Later that evening, he was grading papers with the TV on. He noticed that the winning lottery numbers were being flashed by. He decided to pay attention the next time the numbers were flashed by. When they were, he noticed that all the numbers on one of his lottery tickets agreed with them. As he illustrated for me, he took off his glasses, wiped his eyes, and looked again. By golly, the numbers matched. Though he knew better, he couldn't keep from going into the bedroom, where his wife was sleeping, and waking her. He said, "I think we've won the lottery." She turned over and pulled the blanket over her head. Later, when he couldn't sleep and she turned over, he said, "I think we've won the lottery." She barked, "We'll talk about it in the morning!"

When my friend persisted in this nonsense the next morning, his wife finally decided to end the matter once-and-for-all and said, "Okay, let me see those lottery tickets." When she saw them, she said, "But, the numbers are not in the same order." (For the uninformed, as I was: The order of the "white" numbers does not matter.) Finally, they agreed that they had won the lottery, but they were both too busy on Thursday to notify the authorities. They did so on Friday morning.

My friend met his class Friday afternoon. Just as the students were filing out of the classroom, he said, "Oh, by the way, did any of you buy lottery tickets as I advised?" A few hands went up. "Well, I did and I won." At first they didn't believe him, but because of his story about his class, the press gave extra big publicity to his winning the lottery.

Let's return to the table on page 39 and see where the numbers came from. I'll leave the calculations in Column 2 to the end, except for the bottom TOTAL, 120,526,770. This is the total number of different ways to select 5 numbers from 53 "white" numbers *and* select 1 number from 42 "red" numbers. This calculation breaks into two parts: the number of ways to select 5 numbers from 53 numbers and the number of ways to select 1 number from 42 numbers. The second calculation is easy: There are exactly 42 ways to select 1 number from 42 numbers. Either you select 1, or you select 2, and so forth.

The problem of finding the number of ways to select 5 numbers from 53 numbers is a special case of the problem of finding the number of ways to select some fixed number, say k, of objects from a larger fixed set with, say n, objects. There's a standard notation for this number, namely $_nC_k$. (Actually this is one of two standard notations. The other one is $\binom{n}{k}$, which is often called a "binomial coefficient.") Right now we're interested in the case $k = 5$ and $n = 53$, so we're interested in $_{53}C_5$. The reason that C is part of the notation $_nC_k$ is that selections of objects are sometimes called *combinations* so that $_nC_k$ represents "the number of combinations of k elements from an n-element set." It is often read "n choose k."

Incidentally, you can find $_nC_k$ on some calculators. Fortunately, there's a nice formula for $_nC_k$. Rather than write it out in general, I will describe it in words and illustrate it with some examples. I will discuss why it works at the very end of this chapter. The symbol $_nC_k$ represents the fraction where the denominator is the product of k numbers from k down to 1, and the numerator is the product of numbers starting with n and going down until you also have k numbers in the numerator. Right now we want

$$_{53}C_5 = \frac{53 \times 52 \times 51 \times 50 \times 49}{5 \times 4 \times 3 \times 2 \times 1} = 2,869,685.$$

If you want to check that I correctly followed my own instructions, reread the sentence before the last sentence using $n = 53$ and $k = 5$.

Finally, when counting the number of ways to do two separate things (like selecting 5 balls from 53 white balls *and* selecting 1 ball from 42 red balls), you multiply the number of ways to do each thing. In our case, then, the number of ways of selecting those six numbers in Powerball is

$$_{53}C_5 \times 42 = 2,869,685 \times 42 = 120,526,770.$$

This is the number at the bottom of Column 2 of the table.

I will now explain the number just above 120,526,770 in the table. Recall that nine of the events are ways to win and "None of the above" corresponds to losing. Exactly one of these ten events will occur, and no two of them can occur simultaneously. So the sum of Column 2 needs to be the TOTAL 120,526,770. To get 117,184,724, the number of ways to lose, you simply sum the number of ways of winning and subtract the result from 120,526,770.

All 120,526,770 possible selections in Powerball are equally likely, so the probability of any event is

$$\frac{\text{Number of possibilities of that event}}{120,526,770}.$$

These are the numbers in Column 3. For example, the probability of "3 white balls" is

$$\frac{462,480}{120,526,770} \approx 0.003837.$$

The numbers in Column 4 are provided by the house, that is, by the people who run Powerball. The numbers in Column 5 are the products of the corresponding numbers in Columns 3 and 4. For example, for "3 white balls," this product is

$$\$0.003837 \times 7 \approx \$0.0269.$$

Recall that I provided the numbers in Column 5 because the sum of these products is the expectation for a player who buys one $1 ticket. The product for the jackpot event "all 6 numbers," where the payoff is J million dollars, is

$$\$\frac{J \times 1,000,000}{120,526,770} = \$\frac{J}{120.526770} \approx \$\frac{J}{120.5}.$$

The last column gives the odds of the events. Each of these numbers was obtained by approximating the probability of the event as a fraction $\frac{1}{n}$ for some whole number n and then converting to the odds $1 : (n - 1)$ (this is the Probability to Odds rule on page 24 with $m = 1$). I found n by taking the reciprocal of the probability. For example, the probability of "3 white balls" is .003837 and its reciprocal is

$$\frac{1}{.003837} \approx 260.62 \approx 261.$$

Thus the probability of "3 white balls" is approximately $\frac{1}{261}$ and the odds are 1:260. The odds for losing were determined by first calculating the odds of winning something. The probability of winning something is about $1 - 0.9723 = 0.0277$, because 0.9723 is the probability of "None of the above." Since

$$\frac{1}{.0277} \approx 36.10 \approx 36,$$

the probability of winning something is about $\frac{1}{36}$, so the odds of winning something are about 1:35.

The only thing left to explain are the calculations in Column 2. There's only one way to win, that is, get "all 6 numbers." The event "5 white balls" means that all 5 of the winning "white" numbers were selected and that the winning "red" PB number was not selected. There's only 1 way to select the 5 winning "white" numbers, and there are 41 ways to avoid selecting the winning PB number. Hence there are 1 × 41 ways for the event "5 white balls" to occur.

I will illustrate the rest of the numbers in Column 2 by explaining the ones for "3 white balls & PB" and "3 white balls." The event "3 white balls & PB" means that 3 of the 5 winning "white" numbers were selected, 2 of the 48 losing "white" numbers were selected, and the 1 winning "red" PB number was selected. Because the number of ways to select the winning 3 numbers is $_5C_3 = \frac{5 \times 4 \times 3}{3 \times 2 \times 1}$ and to select the losing 2 numbers is $_{48}C_2 = \frac{48 \times 47}{2 \times 1}$, the total number of ways to get the event "3 white balls & PB" is the product

$$_5C_3 \times {}_{48}C_2 \times 1 = \frac{5 \times 4 \times 3}{3 \times 2 \times 1} \times \frac{48 \times 47}{2 \times 1} = 11,280.$$

The event "3 white balls" means that 3 of the 5 winning "white" numbers were selected, 2 of the 48 losing "white" numbers were selected, and 1 of the 41 losing "red" PB numbers was selected. So there are

$$_5C_3 \times {}_{48}C_2 \times 41 = 11,280 \times 41 = 462,480$$

ways to get the event "3 white balls."

Finally, here is why the formula for $_nC_k$ works, that is, gives the number of ways of selecting k objects from n objects. Consider $_{48}C_2$, the number of ways of selecting 2 objects from 48 objects. You can select the first object in 48 ways and, given that it has been selected, there are 47 ways to select the second object. So there are 48 × 47 ways to select two objects *in order*. But because the order of the objects doesn't matter, this calculation overcounts the number of selections. In fact, since any two objects can be selected in one of 2 orders, 48×47 counts every selection twice. Thus the number $_{48}C_2$ of ways of selecting 2 objects from 48 objects is

$$_{48}C_2 = \frac{48 \times 47}{2} = \frac{48 \times 47}{2 \times 1} = 1128.$$

As one more example, consider $_7C_3$. There are 7 ways to select the first object, 6 ways to select a second object (given that the first object was selected), and 5 ways to select a third object (given that the first two objects were selected). So there are $7 \times 6 \times 5$ ways to select 3 objects *in order*. Suppose *a, b,* and *c* are the names of three objects selected in some order. These 3 objects can be selected in several different orders. The first object can be selected in 3 ways (either *a, b,* or *c*), the second in 2 ways, and the third in 1 way. So there are $3 \times 2 \times 1$ ways to select these 3 objects *in order*. In fact, the orderings are *a b c, a c b, b a c, b c a, c a b,* and *c b a*. This observation holds no matter which 3 objects I'm talking about. Thus $7 \times 6 \times 5$ counts a single unordered selection (of three objects from seven objects) $3 \times 2 \times 1$ times. In other words, $7 \times 6 \times 5$ is $3 \times 2 \times 1$ too big. This is why

$$_7C_3 = \frac{7 \times 6 \times 5}{3 \times 2 \times 1} = 35.$$

The next chapter is concerned with betting on baseball with professionals. To be successful you would want your expectation, namely the expected average dollar gain, to be positive. I will offer a system that may work.

Chapter 4

What Would Pete Rose Do? Professional Betting

I'd walk through hell in a gasoline suit to keep playing baseball.

PETE ROSE

Sports betting, along with such games as poker, is more appealing to many people than standard gambling in casinos. Here's a quote from [28, *Sports Betting 101*, page 12]: "Can the average person earn a living betting sports? No, but neither does one have to be a brain surgeon to come out ahead in the long run. Sports betting appeals to sophisticated gamblers because (1) it is not a negative expectancy game where the house rakes off a fixed percentage of the handle, and (2) sports betting is more a game of skill than a game of luck."

As with any sport, there are many ways to bet on baseball, both in real casinos and in electronic offshore casinos. Straight betting on individual games during the season is my favorite. Here are some typical betting odds offered by offshore casinos:

Seattle	−130,	Minnesota	+120
Yankees	−125,	Cleveland	+115
St. Louis	−210,	Pittsburgh	+190

The betting odds for the Seattle-Minnesota game mean that you can bet on Seattle using odds 130:100 or you can bet on Minnesota using odds 100:120. To bet on Seattle, one risks $130 and hopes to win $100. To bet on Minnesota, one risks $100 and hopes to win $120. You wouldn't want to bet on both. If you did and Seattle won, you'd come out even, and if Minnesota won, you'd lose $10.

My view is that, in essence, the house uses the odds 125:100 and 100:125 and then assesses a "service charge" of $5. If you bet on the favorite, Seattle, the service charge is at the front end: You risk $130 instead of $125. If you bet on the underdog, Minnesota, the service charge is assessed only if you win, in which case you get $120 instead of $125.

If you agree that Seattle is really a 125:100 favorite (i.e., a 5:4 favorite) and if you bet on the game, what is your expectation? If you bet on Seattle, then you risk $130 to gain a $100. Since

$$\mathbf{Pr}(\text{winning the bet}) = \frac{5}{9} \quad \text{and} \quad \mathbf{Pr}(\text{losing the bet}) = \frac{4}{9},$$

your expectation is

$$\frac{5}{9} \times 100 + \frac{4}{9} \times (-130) = -\frac{20}{9} \approx -\$2.22.$$

If you bet on Minnesota, then you risk $100 to win $120 and your expectation is

$$\frac{5}{9} \times (-100) + \frac{4}{9} \times 120 = -\frac{20}{9} \approx -\$2.22$$

again. You will see later that it is not a coincidence that these two expectations are equal. Because the risk is at least $100, the expected average loss per bet in this example is no more than 2.22 percent of the risk.

The good news is that the casino is not charging a very large service charge. The expected average loss per bet is quite moderate. Even in roulette, which we discussed on page 31, the expected average loss per bet is about 5.3 percent of the risk.

It will be convenient to have general notation and formulas so that we don't have to make these calculations over and over. The betting odds that the casinos offer are in the form $a{:}100$ and $100{:}b$. In the Seattle-Minnesota example, $a = 130$ and $b = 120$. Let's write u for the average of a and b, i.e., the midpoint between a and b. The formula is $u = \frac{1}{2} \times (a + b)$. If we let s be half the difference between a and b, that is, if $s = \frac{1}{2} \times (a - b)$, then $u + s = a$ and $u - s = b$. If you mark b, u, and a on a line, you'll see that u is just halfway between b and a, while s is the distance from b to u and also from u to a. In the Seattle-Minnesota example,

$$u = \frac{1}{2} \times (130 + 120) = 125 \quad \text{and}$$

$$s = \frac{1}{2} \times (130 - 120) = 5.$$

Since $u + s = 130$ and $u - s = 120$, the betting odds for Seattle are $(u + s){:}100$ and for Minneapolis they are $100{:}(u - s)$.

In general, the betting odds for the favorite are $(u + s){:}100$ and the betting odds for the underdog are

$100 : (u - s)$. To bet on the favorite, one risks $\$(u + s)$ to win \$100; and to bet on the underdog, one risks \$100 to win $\$(u - s)$. As in the Seattle-Minnesota example with $u = 125$, we can view the odds used by the casino to be u:100 and 100:u, and s has the effect of a "service charge."

You will note that I don't refer to u:100 or 100:u as the "fair" or "true" odds. No one knows what the true odds are, certainly not the house. Then how does the house determine the odds? I'll discuss this later, but you can be sure that the odds are set so that the house cannot lose. The odds offered are close enough to what people perceive as reasonable so as to entice people to gamble.

A Formula

If, in fact, u:100 and 100:u are the true odds, then

$$\text{expectation} = \frac{-s \times 100}{u + 100}.$$

Moreover, the expected average loss per bet is less than or equal to

$$\frac{s \times 100}{u + 100} \text{ percent}$$

of the risk. This is always less than or equal to $\dfrac{s}{2}$ percent.

For the Seattle-Minnesota example, where $u = 125$ and $s = 5$, the formula gives

$$\frac{-5 \times 100}{125 + 100} = -\frac{500}{225} = -\frac{20}{9} \approx -\$2.22$$

for the expectation; this is just what we calculated before.

For the Yankees-Cleveland game, the odds were given as 125:100 for the Yankees and 100:115 for Cleveland. Then $u = \frac{1}{2} \times (125 + 115) = 120$ (this is just the average of 125 and 115) and $s = \frac{1}{2} \times (125 - 115) = 5$ (half the difference between 125 and 115). Using the formula for expectation, if 120:100 and 100:120 are the true odds, then the expectation is

$$\frac{-5 \times 100}{120 + 100} = -\frac{500}{220} = -\frac{25}{11} \approx -\$2.27.$$

As noted after the formula, the expected average loss per bet is at most 2.27 percent of the risk.

For the St. Louis-Pittsburgh game, the casino is offering odds 210:100 for St. Louis and odds 100:190 for Pittsburgh. Then $u = 200$ and $s = 10$. By the formula, if 200:100 and 100:200 are the true odds, then the expectation is

$$\frac{-10 \times 100}{200 + 100} = -\frac{1000}{300} = -\frac{10}{3} \approx -\$3.33.$$

The expected average loss per bet won't exceed 3.33 percent of the risk.

When the teams are evenly matched, or almost evenly matched, the situation is a bit more complicated because both teams will be listed as favorites, to provide for the casino's edge. For one game between the Anaheim Angels and the Seattle Mariners, *both* teams were listed as 105:100 favorites. You could bet $105 on either team, and you'd win $100 if your team won. These odds don't fit the format $u + s:100$ and $100:u - s$. This is because, understandably enough, the casino didn't want to treat one of the teams as the underdog. With the odds 105:100 offered, the expectation is

$$\frac{1}{2} \times 100 + \frac{1}{2} \times (-105) = -\$2.50.$$

I've listed this peculiar evenly matched case at the *end* of the next table.

To avoid the complications, let's ignore the evenly matched cases. Except for these evenly matched cases, one can always use the formula, as I did earlier, but the accompanying table gives the expectations for some commonly offered pairs of odds.

Favorite and Underdog Expectations

Odds for Favorite	Odds for Underdog	The Value u	Service Charge	The Expectation
110:100	100:100	105	$5	-$2.44
120:100	100:110	115	$5	-$2.33
125:100	100:115	120	$5	-$2.27
130:100	100:120	125	$5	-$2.22
140:100	100:125	132.5	$7.50	-$3.23
170:100	100:155	162.5	$7.50	-$2.86
190:100	100:170	180	$10	-$3.57
210:100	100:190	200	$10	-$3.33
220:100	100:195	207.5	$12.50	-$4.07
250:100	100:220	235	$15	-$4.48
300:100	100:260	280	$20	-$5.26
400:100	100:325	362.5	$37.50	-$8.11
105:100	105:100	100	$5	-$2.50

In the preceding examples except for the evenly matched case, for some u bigger than 100 and suitable s, the betting odds for the favorite are given as $(u + s):100$ and the betting odds for the underdog are given as $100:(u - s)$. We view s as the "service charge," and we have been pretending that the odds $u:100$ and $100:u$ are the true odds. These odds are not the *true* odds, because the true odds are simply unknown. The situation is too complicated to assess the true odds, but baseball statisticians and baseball fans, like you, can analyze the situation and come up with their own estimates for the true odds. This is similar to results from polls. Pollsters never know the exact proportion of voters who favor some outcome, but they have methods for estimating the proportions.

The odds $u:100$ and $100:u$ are the ones I would believe if I knew nothing except for the odds given by the house. In general, to make money against the house, you need to know enough about baseball to be confident that the true odds differ enough from these odds to make it possible to overcome the house's advantage. The following betting strategies are pretty obvious, but I will verify below that, if you follow them and if your your judgment is always accurate, then your expectations will always be positive. This means you will win in the long term.

Betting Strategies

- If, in your judgment, the true odds $t:100$ for the favorite are better than the offered odds $(u + s):100$, then bet on the favorite.
- If, in your judgment, the true odds $t:100$ for the favorite are worse than the offered odds $(u - s):100$, then bet on the underdog.
- Otherwise, don't bet.

The "don't bet" line reflects the fact that your belief is too close to that of the house. The little edge you might have is swamped by the service charge. Starting on page 61, I will offer a possible winning system that requires very little knowledge about baseball.

To explain the math behind the formulas and betting strategies, I am going to deal with the case of the underdogs first, because the mathematics is simpler.

Suppose that you bet on the underdog based on the house odds $100:(u - s)$ and you believe the true odds are $100:u$. Then

$$\mathbf{Pr}(\text{winning the bet}) = \frac{100}{100 + u},$$
$$\mathbf{Pr}(\text{losing the bet}) = \frac{u}{100 + u},$$

and your expectation is $\mathbf{Pr}(\text{winning}) \times \text{gain} + \mathbf{Pr}(\text{losing}) \times \text{loss}$, which is

$$\frac{100}{100 + u} \times (u - s) + \frac{u}{100 + u} \times (-100)$$
$$= \frac{100}{100 + u} \times [u - s - u] = \frac{-s \times 100}{100 + u}.$$

This is negative, as you'd expect, and you are best off when the service charge s is small and the value u is large. So the smaller the service charge compared to what you bet, the better off you are. No surprise here. On the baseball gambling sites that I've examined, as u increases the service charge s increases. When this happens, use the preceding formula, or the table on page 54, to determine what's best from the point of view of long-term expectation. Because $\frac{-s \times 100}{100 + u}$ is the expectation based on a $100 bet, the expected loss is the fraction $\frac{s}{100 + u}$ of the risk. Because u is bigger than 100, $100 + u$ is bigger than 200. Hence

$$\frac{100 \times s}{100 + u} \leqslant \frac{100 \times s}{200} = \frac{s}{2},$$

so the expected loss is never larger than $s/2$.

If you bet on the favorite based on house odds $(u + s):100$ and you believe the true odds are $u:100$, your expectation is

$$\frac{u}{u + 100} \times 100 + \frac{100}{u + 100} \times [-(u + s)]$$
$$= \frac{100}{u + 100} \times [u - (u + s)] = \frac{-s \times 100}{u + 100}.$$

This is the same expected loss as when you bet on the underdog. However, the risk is bigger in this case, so the expected loss will be a smaller fraction of the risk. Specifically, since the risk is $u + s$, the expected loss is the following fraction of the risk:

$$\frac{1}{u + s} \times \frac{100 \times s}{u + 100}.$$

Because the expected loss is a smaller fraction of the risk than when betting on the underdog, other things being equal it seems better to bet on the favorite. However, I don't believe that all things are equal in this situation. In fact, my alleged winning system involves betting on certain underdogs.

The preceding computations assume that the odds $u:100$ and $100:u$ are the true odds, so the focus was on the expected loss. Most people would rather win in the long run. The rest of this chapter assumes that the bettor has reason, based on his knowledge or whatever, to believe that the true odds are not the ones used by the casinos.

If the bettor believes the true odds t:100 for the favorite to win are better than $(u + s)$:100, then betting on the favorite is a good betting strategy. This should be obvious, but here's a probabilistic argument why. If he believes this, then he believes that the probability p for the favorite to win is bigger than $\frac{u + s}{u + s + 100}$. I will argue that in this case, his (long-range) expectation is positive. This is because *his* expectation would be

$$p \times 100 + (1 - p) \times [-(u + s)] =$$
$$p \times [100 + (u + s)] - (u + s).$$

The last equality is checked using a little algebra. The last quantity will be positive provided $p \times [100 + u + s] > u + s$, which is true because, as noted earlier,

$$p > \frac{u + s}{u + s + 100}.$$

Now for the underdog. If the bettor believes that the true odds for the underdog to win are better than 100:$(u - s)$, then betting on the underdog is a good betting strategy. The algebra is a little simpler than in the earlier case. With this belief, the bettor believes that the probability q that the underdog will win is bigger than $\frac{100}{100 + u - s}$, and this will make the expectation

$$q \times (u - s) + (1 - q) \times (-100) =$$
$$q \times (u - s + 100) - 100$$

positive.

Finally, note that the gambler favoring the underdog (in the previous paragraph) believes that the true probability p for the favorite to win is less than $\dfrac{u - s}{100 + u - s}$. To see this, first note that $p = 1 - q$. Because the gambler believes $q > \dfrac{100}{100 + u - s}$, he believes $-q < -\dfrac{100}{100 + u - s}$, so he believes

$$p = 1 - q < 1 - \frac{100}{100 + u - s} = \frac{u - s}{100 + u - s}.$$

In terms of p, here is a summary of good betting strategies:

Let p be your personal best assessment of the true probability that the favorite will win. This assessment will likely be based on your knowledge of baseball in general and the teams involved, plus any intuition or hunches you choose to add to the mix.

If $p > \dfrac{u + s}{u + s + 100}$, bet on the favorite.

If $\dfrac{u - s}{u - s + 100} < p < \dfrac{u + s}{u + s + 100}$, don't bet.

If $p < \dfrac{u - s}{u - s + 100}$, bet on the underdog.

These strategies are stated on page 55 in terms of odds.

Up until now, I've discussed betting against professionals who necessarily need to assess "service charges." Now I will explain why the odds u:100 and 100:u are not necessarily the true odds, by which I might mean the average expert opinion. Does each house have a panel of well-paid baseball experts who provide daily odds based on their wisdom and experience? Well, no. In fact, the house does not need any baseball expertise at all. The house bases the odds on how much is bet on the favorite and on the underdog. In other words, it bases the odds on the collective wisdom of the bettors.

To illustrate the idea, suppose that $200,000 is bet on the favorite and $100,000 is bet on the underdog. Let's first assume that there is no service charge, i.e., no vigorish. Then the only safe odds for the house to offer are 1:2 and 2:1 for the underdog and favorite, respectively. Then, if the underdog wins, the $200,000 pile of money is used to pay off the people who bet on the underdog, and if the favorite wins, the $100,000 pile of money is used to pay off the people who bet on the favorite. Either way, the house comes out even.

But the house needs to make a profit. With these amounts of money bet, the house would probably offer odds like 100:185 and 215:100 for the underdog and favorite, respectively. This builds in a service charge of $s = 15$ where $u = 200$. In this case, either the $200,000 pile is used to pay off $185,000 to the people who bet on the underdog, leaving $15,000 for the house, or the $100,000 pile is used to pay $93,023 to the people who bet on the favorite, leaving $6977. (The possibly mysterious $93,023 is 100/215 times $200,000.) Either way, the service charge is 7.5-8 percent of the money transferred.

Just before the last example, I referred to the collective wisdom of bettors. As the example illustrates, this collective wisdom is determined by the dollars bet, not by the number of bettors for each team. In other words, it's "one dollar—one vote," not "one man—one vote." This is similar to the situation in corporations: It's not the number of shareholders who favor a motion that matters; it's the number of shares they own.

Arne Lang's book [28, *Sports Betting 101*] points out that sports betting is "more of a game of skill than a game of luck." This book devotes a few pages to general advice on betting on baseball games (pp. 85–93). For example, "teams that lose a disproportionate number of one-run games are due to improve the next year." Not surprisingly, "age composition of a team's starting lineup was an excellent predictor," with very young teams improving from season to season. A 1982–1984 study by the author Lang shows that "one has the best chance of showing a profit if he zeroes in on small favorites and home underdogs."

Using skill, as suggested above, can be a lot of fun, but I will propose a system for those who want a system that automatically works without having to be on top of the baseball scene on a daily basis. And I don't feel foolish yet even though James Howell observed in 1629: "A fool and his money is soon parted." On the other hand, the pitcher Bill Lee made the following point: "They say that a fool and his money are soon parted. I'd like to know how they got together in the first place."

The System

Bet on weak teams with small fan bases when they play stronger teams with much larger fan bases. In what follows, "much larger" will mean "at least 3 times as large."

I cannot prove that this system works, but I will offer my rationale, and then I will describe my fairly substantial experiment. I decided in advance that I would report the result whether the experiment confirms my theory or not. Feel free to skip my rationale, read about the experiment on page 65, and then read about how it worked out.

My Rationale (March 2003)

Here was my thinking. At each step in my thinking I made simplistic assumptions, but this was my way of guessing a general trend. Only time will tell. That's why I did the experiment over the summer of 2003.

Suppose a strong team with a large fan base, say the Yankees, is playing a weak team with a small fan base, say the Kansas City Royals. (Royals fans, please remember that I wrote this *before* the 2003 season started.) Suppose that the casino offers odds 215:100 for the Yankees and odds 100:185 for the Royals. So $u = 200$ and $s = 15$, where s is the "service charge," so that u:100 and 100:u are the effective fair odds offered by the casino. The true odds t:100 for the Yankees are never known for sure, but I suspect that they are somewhat less than u:100, because the Yankees have more fans who would be inclined to support them than the Royals have who would support them. Below I will make some wild quantitative assumptions along these lines and use them to obtain a formula for t. My belief is that the wild assumptions reflect reality and that, in fact, my assumptions are rather conservative. With these assumptions, t turns out to be about 179.8, so the true odds for the Yankees may be closer to 180:100. Because 180 is less than $u - s = 185$, one should bet on the Royals. Moreover, if the true t is close to 179.8, the expectation turns out to be about $1.86, which is positive!

Here are the assumptions and formulas. The idea is to propose plausible hypotheses under which the expectation would become positive. On a particular day, I will assume that 50 percent of the bettors will be fans of the two teams involved, i.e., either the favorite or the underdog. I assume that the collective wisdom of the bettors, i.e., the money bet for and against the favorite, corresponds to the true odds when the bettors are not biased by being fans of the favorite or underdog. I further assume that the fan base for the favorite is three times larger than the fan base for the underdog; if the ratio is even bigger, the results will only be more pleasing. Finally, and this is the crux of my argument, I assume that the bettors are sufficiently overly enthusiastic for their team that 15 percent more money will be bet on their team than the fractions $\frac{t}{t + 100}$ and $\frac{100}{t + 100}$ would warrant. I cannot prove that this bold assumption is warranted. The only way to test it is to conduct experiments.

Under these **wild** assumptions,

$$t = \frac{100}{103.75 + 0.0375 \times u} \times u,$$

and, *if one bets on the underdog,* the expectation is

$$u - s - p \times [100 + (u - s)], \quad \text{where} \quad p = \frac{t}{t + 100}.$$

Warning: This is not the expectation, under my wild assumptions, if one bets on the favorite. That expectation will be negative. Even if this system works, the house is not in trouble.

I will leave out the calculations, because this is based on wild assumptions, and just give a couple of hints. The assumptions give that the proportion of actual money bet on the favorite would be about

$$\frac{u}{u + 100} = \frac{1}{2} \times \frac{t}{t + 100} + \frac{3}{8} \times \frac{t}{t + 100} \times 1.15$$
$$+ \frac{1}{8} \times \frac{t}{t + 100} \times 0.85. \quad (*)$$

One can solve for t in this equation and then use it to calculate the expectation if one bets on the underdog:

$$\frac{t}{t + 100} \times (-100) + \frac{100}{t + 100} \times (u - s).$$

Look at equation (*). The left side gives the proportion of money bet on the favorite. On the right side, the first product accounts for the proportion of money bet on the favorite by bettors who are not fans of either team, the second product accounts for the proportion of money bet on the favorite by fans of the favorite (1.15 reflects the extra 15 percent bet on the favorite), and the last product accounts for the proportion of money bet on the favorite by fans of the underdog (0.85 reflects that 15 percent less will be bet on the favorite).

Now we show where we got the numbers in the Yankees-Royals example. Since $u = 200$,

$$t = \frac{100}{103.75 + 0.0375 \times 200} \times 200 \approx 179.8.$$

Now $p = \dfrac{t}{t + 100} = \dfrac{179.8}{279.8} \approx 0.6426$. If $s = 15$ and one bets on the Royals, then the expectation is about

$$185 - 0.6426 \times [100 + 185] \approx \$1.86.$$

Consider again the Yankees-Cleveland contest where the casinos were giving odds 125:100 in favor of the Yankees and 100:115 in favor of Cleveland. Here $u = 120$ and $s = 5$. So

$$t = \frac{100}{103.75 + 0.0375 \times 120} \times 120 \approx 110.9$$

and $p = \frac{110.9}{210.9} \approx 0.526$. If one bets on Cleveland, the expectation is about

$$115 - 0.526 \times [100 + 115] \approx \$1.91.$$

Don't forget that these optimistic positive expectations are based on my wild assumptions!

The Experiment

This was a well-designed experiment because I determined in advance, in March 2003, the exact procedure that I would follow. Sometimes people collect data and search for patterns that please them. This is called "mining the data." The trouble is that the patterns they observe might have occurred by chance, because some apparent patterns are bound to occur in any set of data. One can't be sure whether such patterns will occur regularly. This is why I developed my theory *before* I had a chance to look at the data. In case you've forgotten, my theory is that my system will be a winning system in the long run.

I tracked how I would do if, throughout the regular 2003 season,[1] I always, and only, bet on the underdog when the favorite's fan base was at least three times as large as the underdog's fan base. To estimate relative sizes of fan bases, I used the following numbers for the sizes of the fan bases, based on "adjusted metropolitan area populations," which I further adjusted to take into account old established teams that have a national following including many who grew up in the area and moved away.

---

[1] I did not extend the experiment to include the postseason because the bettor population would undoubtedly be different, so I wouldn't expect my wild assumptions to still hold (if they ever did).

21: New York Yankees. 15: New York Mets. 12: Baltimore, Boston, Chicago Cubs, Chicago White Sox, Los Angeles, Philadelphia. 9: Anaheim, Atlanta, Detroit, St. Louis. 7: Cincinnati, Cleveland, Houston, Pittsburgh, Texas, Toronto. 5: Arizona, Florida, Minnesota, Oakland, San Francisco, Seattle. 4: Colorado, San Diego, Tampa Bay. 3: Kansas City, Milwaukee, Montreal.

Thus I bet on Cleveland if Cleveland was the underdog in a Yankees-Cleveland contest because 21 is three times 7, I bet on Milwaukee if Milwaukee was the underdog in a Philadelphia-Milwaukee contest because 12 is more than three times 3, and so forth. I tracked the daily casino odds for all the baseball games using the offshore website BetWWTS.com.

How It Worked Out

Here are the promised results of my experiment. I bet $100 on each of 316 games, always on the underdog. The underdog won 124 of those games, namely 39 percent of them, and lost 192 of those games. Amazingly, I came out about as close to even as one could imagine: *$25 in the hole*. This is a success in that, using the formula on page 52 for the expectation of each bet, I should have lost $1042. On the other hand, I had hoped that I had a winning strategy, not just a strategy to come out even. One also can question whether 2003, or any other year, is typical. My fortune certainly had a boost at the beginning of the 2003 season because Kansas City, Montreal, and Florida started out as big underdogs, but were quite successful throughout the season. Of course, each year will have its own surprises. I will run the experiment again for the 2004 season; for the outcome, see my page on *www.pipress.net*.

The fact that my experiment was only a moral victory suggests that my wild assumptions cannot be taken seriously, at least at this point. Here's a possible alternative psychological theory to explain why I didn't do better. Even though I wasn't betting real money, I was psychologically involved. And I must say, it was more pleasant to risk $100 on Tampa Bay, say, and perhaps make $260 (albeit against the odds) than it would have been to risk $300 on the Yankees, say, to win a paltry $100. So perhaps more people bet on underdogs than the true odds would warrant because it's more pleasant and fun.

Chapter 5

Will the Yankees Win if Steinbrenner Is Gone?
Conditional Probabilities

But in this world nothing is certain but death and taxes.

BENJAMIN FRANKLIN (1789)

The key ideas of this chapter are "independence" and "conditional probability." These ideas are very familiar even if you've never heard the terms or seen numbers called conditional probabilities.

Before discussing a baseball example in detail and using actual numbers, the good news is that you intuitively understand "conditional probability." Given a random 50-year-old American male, there is a probability that he will die of lung cancer. If you have more information, say you're given that he smokes two packs of cigarettes a day, then there would be a different, undoubtedly bigger, probability that he will die of lung cancer. If we had our notation, I'd write

$$\mathbf{Pr}(D|S) > \mathbf{Pr}(D),$$

where

$$D = \text{"he will die of lung cancer,"}$$

$$S = \text{"he smokes two packs of cigarettes a day,"}$$

and the left-side is read "probability of D *given* S." This is shorthand for "the probability of the event D given that the event S has occurred." In a different setting, one might read this as "the probability of D given the condition S," which is why $\mathbf{Pr}(D|S)$ is called a **conditional probability.** Note that if we're told "Smokers are three times as likely as the general population to die of heart disease," we're being told that

$$\mathbf{Pr}(\text{heart disease}|\text{smoker}) \approx 3 \times \mathbf{Pr}(\text{heart disease})$$

or $\mathbf{Pr}(HD|S) \approx 3 \times \mathbf{Pr}(HD)$. If we're told "Smokers are three times as likely as nonsmokers to die of heart disease," then

$$\mathbf{Pr}(\text{heart disease}|\text{smoker})$$
$$\approx 3 \times \mathbf{Pr}(\text{heart disease}|\text{not a smoker})$$

or $\mathbf{Pr}(HD|S) \approx 3 \times \mathbf{Pr}(HD|\text{not } S)$.

Here is another example. If you are a poker or bridge player, you have at least an intuitive idea of the probability that you will be dealt two aces. If, however, you had the additional information that one of your opponents has already been dealt an ace, then you would know that the likelihood that you will get two aces has dropped. Your revised probability is a conditional probability based on extra information.

One more example. Suppose that you are watching a basketball game, that the game is tied with two seconds left in the game, and that you know that a player on your team has just been fouled and will be allowed one foul shot, say. If you know the team well, you'll quickly estimate the probability that the foul shot will be made and hence that your team will win the game immediately. Then you realize that the player who was fouled is your team's worst foul shooter. You'd mutter something like, "Oh darn," as you lowered your estimate of the probability that your team is about to win the game. Your revised probability is another conditional probability.

In baseball, we are often given a batter's batting average against left-handed pitchers, his batting average against right-handed pitchers, his batting average with men on base, and so on. We tend to interpret these numbers as conditional probabilities. For example, the batting average against left-handed pitchers is viewed as the probability of a hit (during an official at-bat) given the extra information that the opposing pitcher is a left-handed pitcher. These numbers are of necessity based on past performance, but they are the best basis for predicting future performance unless we have even more information—such as the batter's record against the particular pitcher or the fact that the batter has a relevant injury. Because of all these human issues, the numbers are slippery and not always believable. For this reason, even though I would prefer our main example to involve baseball, it will come from the medical world.

Most medical tests are not a hundred percent reliable. Results of the tests modify the probabilities that patients have particular conditions. Here's a simplified but realistic example where we analyze some events connected with a test for AIDS. For each patient in the community under consideration (a county or a school system, say), one event of interest is

$$A = \text{"the patient has AIDS."}$$

Another event of interest is

$$T = \text{"the patient would test positive for AIDS,"}$$

i.e., if tested the test would suggest that the patient has AIDS. This particular test is fairly reliable in the sense that 95 percent of patients with AIDS would test positive. So it serves as partial confirmation when a patient is suspected of having AIDS. With our new notation, I can write

$$\mathbf{Pr}(T|A) = 0.95.$$

In words, the probability that the patient would test positive given that the patient actually has AIDS is 0.95, i.e., 95 percent of the people with AIDS would test positive. Does it follow that this fairly reliable test should be given to everyone? Or, at least, should a random sample of people from the community be tested? I will return to these questions later.

It turns out, in my story, that studies show that

$$\mathbf{Pr}(A) = 0.01.$$

That is, one percent of the people in the community actually have AIDS. What is the probability that a randomly selected person in the community has AIDS *and* would test positive? In symbols, what is $\mathbf{Pr}(A$ and $T)$? Because the proportion of a proportion is a product of the proportions,

$$\mathbf{Pr}(A \text{ and } T) = \mathbf{Pr}(A) \times \mathbf{Pr}(T|A)$$
$$= 0.01 \times 0.95 = 0.0095.$$

This is reasonable because one percent of the people in the community has AIDS and 95 percent of them would test positive.

The last equation is correct in general, even when the probabilities cannot be calculated using proportions. That is, for any events E and F, we have

$$\mathbf{Pr}(E \text{ and } F) = \mathbf{Pr}(E) \times \mathbf{Pr}(F|E).$$

By dividing both sides by $\mathbf{Pr}(E)$, we get a nice formula for the **conditional probability**:

$$\mathbf{Pr}(F|E) = \frac{\mathbf{Pr}(E \text{ and } F)}{\mathbf{Pr}(E)}.$$

If we interchange E and F, we get another formula:

$$\mathbf{Pr}(E|F) = \frac{\mathbf{Pr}(F \text{ and } E)}{\mathbf{Pr}(F)} = \frac{\mathbf{Pr}(E \text{ and } F)}{\mathbf{Pr}(F)}.$$

These formulas are not the same! As we will see in the next example, $\mathbf{Pr}(F|E)$ and $\mathbf{Pr}(E|F)$ are quite different objects.

I continue with the AIDS example. Recall

$$\mathbf{Pr}(T|A) = 0.95 \quad \text{and} \quad \mathbf{Pr}(A) = 0.01.$$

It turns out that, in my story, studies also show that

$$\mathbf{Pr}(T) = 0.03.$$

In other words, 3 percent of the folks in the community would test positive. I already calculated $\Pr(A \text{ and } T) = 0.0095$. I now have enough information to calculate

$$\Pr(A|T) = \frac{\Pr(A \text{ and } T)}{\Pr(T)} = \frac{0.0095}{0.03} \approx 0.3167.$$

From the point of view of the diagnosis, though not of the patient, the good news is that this is a lot bigger than $\Pr(A)$. The bad news is that $\Pr(A|T)$ is a long way from 1.00. In fact,

$$\Pr(\text{"not } A\text{"}|T) = 1 - \Pr(A|T) \approx 0.6833,$$

because if T occurs either event A occurs or event "not A" occurs. The unpersuaded reader can verify this using a little algebra and the fact that

$$\Pr(T \text{ and } A) + \Pr(T \text{ and not } A) = \Pr(T).$$

In any case, $\Pr(\text{"not } A\text{"}|T) \approx 0.6833$ is very bad news. About 68 percent of randomly selected people who test positive will not have AIDS! This can be misleading and upsetting to both the individuals involved and to society, and the results could be misused or misinterpreted as well. Tests that are fairly reliable in one sense may be considerably less reliable in other senses. Such tests should not be used indiscriminately.

There is a brighter side to the story. Since

$$\begin{aligned}\Pr(A \text{ and not } T) &= \Pr(A) - \Pr(A \text{ and } T) \\ &= 0.01 - 0.0095 = 0.0005\end{aligned}$$

and $\Pr(\text{not } T) = 1 - \Pr(T) = 1 - 0.03 = 0.07,$

we have

$$\Pr(A|\text{not } T) = \frac{\Pr(A \text{ and not } T)}{\Pr(\text{not } T)} = \frac{0.0005}{0.97} \approx 0.0005.$$

So if the test is *negative*, the probability is very low that the individual has AIDS. An individual needs to weigh the possible outcomes before having the test: A positive result might well be misleading, but a negative result would be very reassuring. A concerned individual should probably be tested, but if the result is positive, no one should assume that the individual has AIDS. Further testing and analysis is needed. I should stress that these remarks apply to the test in my story; they won't apply to all tests in the real world.

If you found the calculations of this example too abstract, pretend that the total population of the community is exactly 10,000 and that we know exactly who has AIDS and who would test positive. Let's pretend that the numbers are in the following table:

	A	Not A	Totals
T	95	205	300
Not T	5	9695	9700
Totals	100	9900	

Thus 95 people have AIDS and would test positive, 205 don't have AIDS but would test positive, and so forth. In all, 300 would test positive and 9700 would test negative. With this unrealistically complete data, all the probabilities and conditional probabilities would be the same as before, but now could be calculated as simple proportions. As one example, $\Pr(A|\text{not } T)$ would be

$$\frac{\text{number with AIDS who would test negative}}{\text{number who would test negative}} = \frac{5}{9700}$$

$$\approx 0.0005.$$

I apologize that the preceding examples aren't about baseball. The problem is that any interesting conditional probabilities involve future unknown outcomes, but they need to be based on past performance. Baseball is such a complicated game that the conditional probabilities fast become meaningless. In particular, they would be meaningless by the time this book is published.

In all of our examples so far, the conditional probabilities have been different from the original probabilities, and this is what made the examples interesting. You might guess that the case where the conditional and original probabilities are the same would be of little interest. Actually, this case is very important in probability, because it is often important to know when new information does *not* change the probability of an event. When this occurs, we say that the two events (the original event and the new information) are "independent." We really mean that they are "probabilistically independent," because knowing that one occurs doesn't affect the probability that the other occurs. Knowing that one occurs certainly provides more information, but we regard the events as "independent" if the probabilities are unaffected by the new knowledge.

The alert reader may have noticed that I first described independence in a nonsymmetric way, because only one conditional probability was involved, but then I talked about two events being independent of each other in a symmetric way. It is a little mathematical fact that these two ways of talking about independence are equivalent.

Let's return again to the day when the Atlanta Braves were 3:2 favorites to win their game and the Chicago Cubs were 4:3 favorites to win their game; see pages 18 and 24. We used the notation

$$B = \text{event that the Braves win}$$

and

$$C = \text{event that the Cubs win.}$$

If we knew the outcome of one of the games, there's no reason to believe that this knowledge would change the probability for the outcome of the other game. In other words, we believe the events B and C are "independent." In terms of conditional probabilities, we believe

$$\mathbf{Pr}(B|C) = \mathbf{Pr}(B) \quad \text{and} \quad \mathbf{Pr}(C|B) = \mathbf{Pr}(C).$$

After formalizing the concept of "independence," I will return to this example.

Here's a fundamental fact about conditional probabilities.

Consider events E and F. The following three statements are equivalent. That is, either all of them hold for our particular E and F or else none of them hold.

(a) $\mathbf{Pr}(E|F) = \mathbf{Pr}(E)$.
(b) $\mathbf{Pr}(F|E) = \mathbf{Pr}(F)$.
(c) $\mathbf{Pr}(E \text{ and } F) = \mathbf{Pr}(E) \times \mathbf{Pr}(F)$.

This is true because (a) means

$$\frac{\mathbf{Pr}(E \text{ and } F)}{\mathbf{Pr}(F)} = \mathbf{Pr}(E)$$

and (b) means

$$\frac{\Pr(F \text{ and } E)}{\Pr(E)} = \Pr(F),$$

so all three equations say the same thing.

We say that the events E and F are **independent** if any of the three properties (a), (b), or (c) hold, in which case they all hold. Because (a) and (b) either both hold or both fail to hold, there's no ambiguity whether we say, "E is independent of F," "F is independent of E," or "E and F are independent." So the alert reader was correct in noting that I was mixing symmetric and nonsymmetric descriptions of independence, but I was okay because the descriptions are equivalent.

Because we believe the outcomes of the Braves' and Cubs' games are independent, property (c) tells us that

$$\Pr(B \text{ and } C) = \Pr(B) \times \Pr(C).$$

I used this equation on page 25 when I was explaining the rather strange-looking odds, 29:6 and 12:23, claimed on page 18.

Warning: The product formula $\Pr(E \text{ and } F) = \Pr(E) \times \Pr(F)$, which is nice and very tempting, does *not* work in general. But it works whenever the events are independent.

I am in a nationwide long-term study to determine whether vitamin E or silenium help reduce the chances of getting prostate cancer. One important assumption in the study is that the outcomes for the 32,000 or so people in the study are independent. Surely they are. There's no reason to believe that, if I ended up getting prostate cancer, this would change the likelihood that others would get it.

On the other hand, suppose that a large number of people are studied to test the effectiveness of a cold remedy. To save money, the study is restricted to a small area, like New York City or Rhode Island. Would this be a good study? No. The outcomes of the individuals are not likely to be independent, because colds are contagious. If you have a cold, this increases the likelihood that others near you will have a cold.

Independence is arguably the most important concept in probability. It is the concept that distinguishes probability from other areas of mathematics. I've heard mathematicians say that probability is just a special case of their areas, but they are ignoring the fact that independence adds a flavor to the subject missing in those other areas.

In the next chapter independence will play a key role in the study of what happens when events can be repeated many times. One should always question whether independence is a reasonable assumption. In the story about the Braves' and Cubs' games, it is reasonable to assume that the outcomes are independent *unless* something very unusual is going on. For example, if the games were being played on the last day of the season, if the Braves were tied with the Phillies for the division title, and if the Cubs were playing the Phillies earlier in the day than the Braves' game, then I would no longer believe the outcomes of the games are independent, because the outcome of the Cubs-Phillies game would surely have an effect on the players, manager, and coaches in the Braves' game.

Chapter 6

How Long Should the World Series Last?
Repeated Tries

If the World Series goes seven games, it will be NBC's longest running show this fall.

JOHNNY CARSON, TALK SHOW HOST (1978)

Next I focus on what happens if the same event occurs over and over with the same probability. A typical question, involving a ballplayer with a known batting average, is "How likely is it that he will get two hits in his next four official at-bats?" Certain assumptions will be made and they have their limitations.

Recall the roulette example on page 31, where we saw that the odds are 9:10 that you'll win one game of roulette if you bet on red. If the machinery is in good working order and the casino is honest, these odds are fixed and won't change from one game to the next. The probability p of winning a game is $\frac{9}{9+10} = \frac{9}{19} \approx 0.4737$, by the Odds to Probability rule on page 23. If you choose to play roulette many times, always betting on red, note that

(a) Each time there are two possible outcomes, winning or losing.

(b) The probability p of winning is the same each time.

(c) The events are independent.

Let's clarify statement (c): The events are independent. The events that I'm referring to are the individual games, i.e., the first game, the second game, etc. By **independence** here, I mean that the outcome of a particular game is (probabilistically) independent of the outcomes of some or all of the other games. Knowing that you won the first and third games does not change the probability that you won, or will win, the fourth game.

When some event is repeated over and over, such as playing roulette or taking an at-bat in baseball, it is always a fair question whether the events are independent. With roulette this is certainly a reasonable assumption. With other gambling devices, like slot machines, it is again a reasonable assumption if the casino is honest. Note, however, that with slot machines there are many possible outcomes of interest, so property (a) in the first paragraph does not hold. Similar comments apply to video poker. What about in sports, where athletes repeat an event over and over?

Before discussing the situation in baseball, let's consider basketball where some of the distinctions are easier to see.

Basketball players score points by making field goals and by making free throws. A good player's field goal percentage might be 60 percent and her free throw percentage might be 90 percent. This means that, over the long term, it has been observed that she makes about 60 percent of her field goal attempts and about 90 percent of her free throw attempts.

Let's discuss field goals first. That is, let's analyze a player's sequence of field goal *attempts*. These are the events and there are two possible outcomes, success or failure. Is it true that

(a) Each time there are two possible outcomes, success or failure?
(b) There's a fixed probability p of success each time?
(c) The events are independent?

Well, I just said that (a) holds; that's easy. However, even though she's a 60 percent shooter, it isn't reasonable to assert that (b) holds with, say, $p = 0.60$. Sometimes she takes difficult shots and attempts three-pointers; other times her attempts are relatively easy. The skills of her opponents will also vary. The situation is definitely more complicated than property (b) would suggest. Property (c) won't be totally true either. Based on earlier performance, her coach might give her new instructions, the opposing team might adjust its strategy, her self-confidence or health might change, and so on. Properties (a), (b), and (c) will not generally hold for field goal attempts.

Now let's discuss a sequence of free throws. I claim that assumptions (a), (b), and (c) are reasonable assumptions with $p = 0.90$. Though human beings are not robots, and hence are subject to different pressures, in this case I believe it is reasonable to assume properties (b) and (c) are approximately true. They won't be certifiably correct assumptions, but should be close to the truth.

Bowling and horseshoes are sports activities where the assumptions (a), (b), and (c) are even more reasonable than with free throws. But let's return to our favorite subject.

Consider some ballplayer and a string of official at-bats. I will focus on two possible outcomes for an official at-bat, a hit (success) or not (failure). Using either a short-term or long-term batting average, I get a probability p

that probably represents the *average* likelihood that he will get a hit in future official at-bats. But we have very little reason to believe that p is even close to the correct probability of success for *each* official at-bat. Indeed, it's difficult to assign any value to the probability of a hit in a particular official at-bat, but it is clear that it will change a lot depending on the opposing pitcher, on whether the pitcher is left-handed or right-handed, on instructions from the manager, on the defense the batter is facing, on the weather and playing field, etc. I'm sure that you can add to this list. So the situation is very complex and properties (a), (b), and (c) do not all hold, even if I could make a case that the separate official at-bats are independent events.

Because baseball is a very sophisticated game involving many players and other complicating factors, I am unable to think of any repeatable baseball event that approximately satisfies properties (a), (b), and (c) like free throw attempts do in basketball.

In spite of these discouraging comments, it is often extremely useful to *pretend* that properties (a), (b), and (c) hold, and then compare this idealized situation with the real-world situation. You can think of this as comparing a robot, programmed to always behave in the same way, with very real human beings. When people ask questions like, "Do baseball players have hitting streaks?," they mean "as compared with our idealized robot." Because of the nature of random (probabilistic) behavior, even a robot whose hitting satisfies properties (a), (b), and (c) precisely would experience streaks of hits. So the question posed earlier is really, "Do baseball players tend to have more and longer hitting streaks than they would if they were robots (with the same probability p of success)?" Right now we're looking at probabilities based on some assumptions.

When we get to statistics in Chapter 8, where I will return to the question of streaks in sports, we'll see how data can be analyzed to determine whether our assumptions are reasonable.

Repeatable situations satisfying properties (a), (b), and (c) are so important that they have a name. Suppose we have a sequence of events satisfying

(a) Each time there are two possible outcomes, traditionally called *success* and *failure*.

(b) There is a fixed probability p of success each time.

(c) The events are independent.

In general, the events satisfying (a), (b), and (c) are called **Bernoulli trials.** They are called "trials" (meaning "tests" or "experiments") because of important applications to medicine, agriculture, and other fields. They are called "Bernoulli" trials in honor of the great Swiss mathematician Jakob (or Jacques or James, all were used) Bernoulli (1654–1705) who did some early work on this subject at the dawn of the scientific revolution. Jakob was the first of several Bernoullis who were great mathematicians, just as there was a whole family of Bachs who were great musicians.

Bernoulli trials are important for several reasons. They serve as a model for many situations. Sometimes they are a perfect model (such as in the roulette example), sometimes they are a pretty good model (such as in the free throw example), and sometimes they aren't such good models. But, even when they are a poor model, they can serve as a benchmark with which to compare reality. Another good thing about Bernoulli trials is that they are very well understood from the point of view of probability theory. In this book, I am just scratching the surface.

To state our key result, we need the notation $_nC_k$ from Chapter 3; see page 43. Recall that $_nC_k$ is the number of ways to select k objects from a set of n objects. For example, $_7C_3$ is the number of ways to select 3 objects from 7 objects. In fact,

$$_7C_3 = \frac{7 \times 6 \times 5}{3 \times 2 \times 1} = 35.$$

Key Results

Given a sequence of n Bernoulli trials, with probability p of success at each trial and probability q of failure at each trial (so $p + q = 1$),

$$\mathbf{Pr}(\text{exactly } k \text{ successes in } n \text{ trials}) = {_nC_k} \times p^k \times q^{n-k}.$$

Also, the expected number of successes is $n \times p$.

The last sentence means that if the entire experiment of n trials could be repeated many times, then the average number of successes over all these experiments would be close to $n \times p$.

Let's return to the roulette example and assume that you bet on red 100 times. For each bet, the probability of success is $p \approx 0.4737$. Because we assume that casinos are honest, this is a bona fide sequence of Bernoulli trials with $p = 0.4737$ and $n = 100$.

As stated in the Key Results, the expected number of successes is $n \times p = 100 \times 0.4737 = 47.37 \approx 47$. On average, you'd expect to win about 47 times out of 100 games. No surprise here.

The formula for $\mathbf{Pr}(\text{exactly } k \text{ successes})$ is much less intuitive. For large n, like 100, it is also difficult to calculate and not very useful by itself. As an example, consider the probability of *exactly* 47 wins in betting on red in roulette 100 times. Since $q = 1 - p = 0.5263$, by the Key Results, this is

$$\text{Pr}(\text{exactly 47 successes in 100 trials})$$
$$= {}_{100}C_{47} \times (.4737)^{47} \times (.5263)^{53}.$$

Arrghh!! Both of the powers $(.4737)^{47}$ and $(.5263)^{53}$ are extremely small, and ${}_{100}C_{47}$ is huge and unmanageable. The numerator of ${}_{100}C_{47}$ is the product of 47 numbers, $100 \times 99 \times 98 \times \cdots \times 54$, and the denominator is also a product of 47 numbers, $47 \times 46 \times 45 \times \cdots \times 2 \times 1$. Even though I love my hand-held calculator, I give up.

Fortunately, when n is large (say, bigger than 15 or so), people don't care about these numbers, because each of them is quite small. They care about the probability that the number of successes lies in some interval. In our example, the probability

$$\text{Pr}(\text{number of successes is bigger than 37} \qquad (*)$$
$$\text{and less than 57})$$

would be quite interesting. Note that 47 is midway between 37 and 57. This is the sum of 19 of those unmanageable numbers, so this is even harder to calculate than $\text{Pr}(\text{exactly 47 successes})$. But it turns out that there are effective formulas and tables that make it possible to approximate the probability in $(*)$. In our case, the probability using the DeMoivre-Laplace Limit Theorem is approximately 0.942. In this experiment of betting on red 100 times, there is almost a 95 percent chance that you'll win more than 37 times and less than 57 times. If you won more than 56 times, you were extremely lucky. If you won less than 38 times, you might wonder whether the game was honest—or you were just very unlucky.

We return to the sharpshooting basketball player on page 82. As I indicated there, I believe that her attempted free throws are close to Bernoulli trials with $p = 0.90$. Does this mean that she is sure to make *exactly* 9 of her next 10 free throws? Nope! The Key Results tell us that the probability that she'll make exactly 9 of her next 10 free throws is about

$$_{10}C_9 \times (.90)^9 \times (.10)^1.$$

Recall that $_{10}C_9 = {}_{10}C_1 = 10$, so this probability is about

$$10 \times (.90)^9 \times \frac{1}{10} = (.90)^9 \approx 0.387.$$

The probability that she'll make exactly 8 of her next 10 free throws is

$$_{10}C_8 \times (.90)^8 \times (.10)^2.$$

Since $_{10}C_8 = {}_{10}C_2 = \frac{10 \times 9}{2 \times 1} = 45$, this is $45 \times (.90)^8 \times (.10)^2 \approx 0.194$. All the probabilities of interest are given in the next table, each to three decimal places. They were all calculated using the Key Results. Note that the sum of the probabilities is 1.000, as it should be. The prob-

Probability of k Successes in 10 Tries, $p = 0.90$

k	10	9	8	7	6	5	0-4
Pr(exactly k successes)	.349	.387	.194	.058	.011	.001	≈..000

ability that she'll make at least 8 of the free throws is the sum of the three numbers **Pr**(exactly 8 successes), **Pr**(exactly 9 successes), and **Pr**(exactly 10 successes), so

Pr(at least 8 successes) = .194 + .387 + .349 = .930.

It's interesting to observe that

Pr(at most 8 successes) = 1 − (.349 + .387) = .264,

so she is a lot more likely to make all 10 free throws than she is to make less than 9 of them.

Because the probability that she'll make less than 6 of her next 10 free throws is about .001 (very small!), if in fact she did so, it would be reasonable to conclude that these 10 attempts were not Bernoulli trials. Something went wrong! Either her probability p of success at each trial was no longer 0.90, her probability of success at the trials varied, or perhaps the trials weren't independent.

Even though we are less confident that her field goal attempts are even approximately Bernoulli trails, let's do some calculations anyway. We assume Bernoulli trials with $p = 0.60$ and consider the next 5 field goal attempts. She should make 60 percent of them, namely 3, right? Well, no, the probability would be ${}_5C_3 \times (.60)^3 \times (.40)^2$. This is

$$\frac{5 \times 4 \times 3}{3 \times 2 \times 1} \times (.60)^3 \times (.40)^2$$
$$= 10 \times (.60)^3 \times (.40)^2 \approx 0.3456.$$

She has less than a 35 percent chance of having *exactly* 3 successes. The next table gives all the probabilities of interest.

Probability of k Successes in 5 Tries, p = 0.60

k	0	1	2	3	4	5
Pr(exactly k successes)	.0102	.0768	.2304	.3456	.2592	.0778

Note that our basketball player is more likely to have 3 successes than any other fixed number of successes. However, since

$$\text{Pr}(\text{number of successes is } not \; 3) = 1 - 0.3456$$
$$= 0.6544,$$

she's more likely to fail to have 3 successes.

Even though a batter's official at-bats, where "success" is "getting a hit," are not really Bernoulli trials, it is still interesting to pretend that they are and see what the probabilities are. If we tried to be more precise and put different probabilities on each trial and/or use conditional probabilities (i.e., not assume independence), the problem would quickly become impossibly complicated.

In the very first chapter, I asked the following innocent-looking question: How likely is it that a ballplayer will get two hits in his next four official at-bats? This is pretty vague! Did I mean "exactly two hits" or "at least two hits"? As I indicated in the previous paragraph, to have a chance at this problem, I need to pretend we're dealing with Bernoulli trials with $n = 4$. What is p? It depends on the ballplayer. If I don't specify a value of p, then (assuming that the at-bats are Bernoulli trials) the probability that the ballplayer will get exactly two hits in his next four official at-bats is

$$_4C_2 \times p^2 \times q^2 = 6 \times p^2 \times q^2 \quad \text{where} \quad q = 1 - p.$$

If the ballplayer is Barry Bonds and I use his batting average for 2002 for p, then $p = 0.370$ from Chapter 1 and the probability that he will get exactly 2 hits in his next four official at-bats is

$$6 \times (.370)^2 \times (.630)^2 \approx 0.326.$$

The probability that he'll get at least 2 hits is

$$
\begin{aligned}
&_4C_2 \times p^2 \times q^2 + {_4C_3} \times p^3 \times q + {_4C_4} \times p^4 \\
&= 6 \times (.370)^2 \times (.630)^2 \\
&\quad + 4 \times (.370)^4 \times (.630) + (.370)^4 \\
&\approx .326 + .128 + .019 = 0.473.
\end{aligned}
$$

In general, the expected number of successes in n Bernoulli trials, with probability p of success of each trial, is $n \times p$. We have interpreted this as saying that if n is large, then the fraction of successes in n trials will be close to p. So the fraction

$$\frac{\text{number of successes in } n \text{ trials}}{n}$$

will be close to p if n is large. This isn't a guarantee. One could be extremely lucky and have success every time or be extremely unlucky and have no successes. If this isn't a guarantee, then what is true? In ways that can be made mathematically precise, one can be "almost sure" that when n is large then the fraction

$$\frac{\text{number of successes in } n \text{ trials}}{n}$$

is close to p. Results that state this precisely clarify exactly what we mean by "almost sure" and are called "Laws of Large Numbers." These results are important and interesting, but they are also easily misunderstood. For example, it is *not* true that, after a long run of failures, one is "due" a success. Perhaps one is morally "due" a success, but the likelihood of success does not increase just because there's been a string of failures. If this were true, then the events would not be independent and we would not be looking at Bernoulli trials. To see this clearly, think of the case of playing roulette or slot machines where the results are really independent. The machine will not remember whether you've just lost the last 20 games. Of course, with people in the action the situation is always more complicated. If you can convince a ballplayer who is in a slump that he is "due" for a hit, then this might by itself increase his probability of success on his next attempt. This would be good for the ballplayer, but his performance would then deviate from the Bernoulli trial model.

I now explain the Key Results on page 86. We are given n Bernoulli trials with probability p of success on each trial. As usual, $q = 1 - p$ is the probability of failure on each trial. Our first task is to find a formula for

$$\mathbf{Pr}(\text{exactly } k \text{ successes in } n \text{ trials}).$$

I'll illustrate the general argument with an example. Suppose we want to calculate the probability of exactly 4 successes in 11 trials. One way for this to happen would be to have the following sequence of events

$$S\,F\,F\,F\,F\,S\,F\,F\,S\,S\,F,$$

where each S signifies a success and each F signifies a failure. This string is shorthand for the statement, "the first trial was a success, the next four trials were failures, the sixth was a success, and so forth." What is the probability of this exact sequence of independent events? It is the product of the probabilities of each event, namely

$$p \times q \times q \times q \times q \times p \times q \times q \times p \times p \times q$$
$$= p^4 \times q^7.$$

In fact, every sequence of four Ss and seven Fs will have probability $p^4 \times q^7$. Moreover, the probability of exactly 4 successes in 11 trials will be the sum of all these probabilities. Because these probabilities are all the same number, namely $p^4 \times q^7$, we just need to know how many of these numbers we are summing. This is the number of strings of 11 letters with 4 Ss and 7 Fs, which is the number of ways of selecting 4 of the 11 positions for the Ss. This is the number $_{11}C_4$. So the probability of 4 successes in 11 trials is the product

$$_{11}C_4 \times p^4 \times q^7.$$

In general, the probability of k successes in n trials is $_nC_k \times p^k \times q^{n-k}$.

With these probabilities, we can calculate the expected number of successes in n trials. It is the sum of the products

$$\mathbf{Pr}(k \text{ successes in } n \text{ trials}) \times k = {}_nC_k \times p^k \times q^{n-k} \times k,$$

where k takes all the $n + 1$ values from 0 to n. This sum is not pretty, though there are tricks of the trade that make it possible to show that this very complicated sum is simply $n \times p$.

Here's a much more intuitive explanation that can be made into a mathematically precise argument. In a single experiment in a trial, the expected number of successes is $p \times 1 + q \times 0 = p$. A Bernoulli trial with n experiments is the sum of n single experiments, and it turns out that one can just add the expectations. So the expected number of successes in n trials is

$$p + p + \cdots + p \quad (n \text{ times}) = n \times p.$$

I now turn to some questions about World Series, which I mentioned on pages 34 and 35. Because there's no time limit in baseball, even when the outcome of a game seems clear, sometimes defeat can be grasped from the jaws of victory. So there's a special interest in how long the World Series will last. As Yogi Berra has warned us, "it ain't over 'til it's over!"

I will make some assumptions that will ignore important elements of the game like the choices of pitchers and home advantages. In fact, I will assume that the series consists of Bernoulli trials with a fixed probability p which I'll pretend applies to each game:

- Each game will be won by the favored team with probability p.

If the teams are equally favored, then $p = \frac{1}{2}$. Otherwise, p will be bigger than $\frac{1}{2}$.

But what is n? That was the original question: How long will the series last? So this is not exactly a problem of Bernoulli trials, but the problem breaks down into several pieces that are Bernoulli trials. Before I give the explanations, here are the answers, where as usual $q = 1 - p$.

Probabilities for the Lengths of World Series

Length of Series	4	5	6	7
Favored wins	p^4	$4p^4q$	$10p^4q^2$	$20p^4q^3$
Favored loses	q^4	$4pq^4$	$10p^2q^4$	$20p^3q^4$
Totals	$p^4 + q^4$	$4[p^4q + pq^4]$	$10[p^4q^2 + p^2q^4]$	$20p^3q^3$

For example, the probability that the favored team will win the series in five games is $4p^4q$, and the probability that the series will last seven games is $20p^3q^3$. A couple of more comments are needed. I am finally using the notational agreement from algebra that adjacent numbers or letters, without any operation indicated, are to be multiplied. Thus $4p^4q$ is shorthand for $4 \times p^4 \times q$ and $20p^3q^3$ is shorthand for $20 \times p^3 \times q^3$. The totals in the table are clear except that it looks like I made a mistake in the last column. But I didn't, because

$$20p^4q^3 + 20p^3q^4 = 20p^3q^3 \times [p + q]$$
$$= 20p^3q^3 \quad \text{since} \quad p + q = 1.$$

It is a somewhat challenging pure-algebra problem to verify that the four probabilities in the Totals line add to 1, as they should. The argument will be outlined on page 171.

If the teams are evenly matched, then $p = q = 0.50$ and the numbers in the Totals line of the table are $(.5)^4 + (.5)^4 = 0.125$, $4 \times [(.5)^5 + (.5)^5] = 0.250$, $10 \times [(.5)^6 + (.5)^6] = 0.3125$, and $20 \times (.5)^6 = 0.3125$, respectively. These are the numbers used for the probabilities of these events on page 35. Back there I used these probabilities to calculate the expected length of a World Series, which turned out to be 5.8125.

Now let's suppose that the teams are not evenly matched, say $p = 0.60$ and $q = 0.40$. These are the values of p and q if the favored team is a 3:2 favorite in *each game*. This is different from the odds that the favored team will win the entire series; see the next paragraph! Under these assumptions, the probability that the series will go five games is

$$10[p^4q^2 + p^2q^4] = 10[(.6)^4(.4)^2 + (.6)^2(.4)^4]$$
$$= 10[.020736 + .009216] \approx .2995.$$

This and the other probabilities for the length of the series are summarized in the following table. Incidentally, if you stare at the last calculation, you can see that the probability that the series will go five games *and* be won by the favorite is about 0.207, while the probability that the series will go five games *and* be won by the underdog is only about 0.092. Given that the series went five games, the conditional probability that the favored team won the series is about

$$\frac{\text{Pr}(\text{series went five games and was won by favorite})}{\text{Pr}(\text{series went five games})}$$

$$\approx \frac{0.207}{0.2995} \approx 0.69.$$

What is the probability that the favored team will win the World Series? This is not the same as the probability

Length of World Series

Length of Series	4	5	6	7
Probability [$p = 0.5$]	0.125	0.250	0.3125	0.3125
Probability [$p = 0.6$]	0.1552	0.2688	0.2995	0.2765

for winning each game, which we've assumed is 0.60. I need to sum the probabilities in the second row of the table on page 95. I get

$$p^4 \times [1 + 4q + 10q^2 + 20q^3],$$

where $p = 0.6$ and $q = 0.4$. So this probability is

$$(.6)^4 \times [1 + 4 \times (.4) + 10 \times (.4)^2 + 20 \times (.4)^3]$$
$$= (.6)^4 \times [1 + 1.6 + 1.6 + 1.28],$$

which simplifies to $(.6)^4 \times [5.48] \approx 0.71$. This is not a number you could have guessed.

Of course, all these calculations are based on the assumption that we're dealing with Bernoulli trials. This is not a true statement, but is probably close enough to give us reasonable guides as to what to expect. While I am confessing, I might as well admit that the value of p itself is impossible to determine. It's another educated guess.

Here are some typical conditional probabilities that come up during a World Series. Given that a team wins the first game, what is the probability that that team will win the series? What if the team wins the first two games or wins the first three games? The answers are in the next table.

In the next table, I provide the numbers with $p = 0.50$ and with $p = 0.60$. I also provide the actual percentages of

Conditional Probabilities Team Wins Series

Given that a team	Conditional probability that team wins series
wins first game	$p^3 \times [1 + 3q + 6q^2 + 10q^3]$
wins first 2 games	$p^2 \times [1 + 2q + 3q^2 + 4q^3]$
wins first 3 games	$p \times [1 + q + q^2 + q^3]$

series won under these conditions, based on the 94 4-out-of-7 series used earlier.

Specific Conditional Probabilities Team Wins Series

Given that a team	with $p = 0.50$	with $p = 0.60$	Observed
wins first game	≈ 0.656	≈ 0.821	≈ 0.606 (57/94)
wins first 2 games	≈ 0.812	≈ 0.913	≈ 0.783 (36/46)
wins first 3 games	≈ 0.937	≈ 0.974	1.000 (20/20)

The observed percentages don't match the "predicted" ones very well.

Note that **Pr**(series will go at least 6 games) is equal to

$$10[p^4q^2 + p^2q^4] + 20p^3q^3 = 10p^2q^2[p^2 + q^2 + 2pq]$$
$$= 10p^2q^2(p + q)^2.$$

Since $p + q = 1$, this simplifies to

$$\mathbf{Pr}(\text{series will go at least 6 games}) = 10p^2q^2.$$

Now the conditional probability that the series will go exactly 6 games, given that it went at least 6 games, is

$$\frac{\mathbf{Pr}(\text{series will go exactly 6 games})}{\mathbf{Pr}(\text{series went at least 6 games})},$$

and this is equal to

$$\frac{10[p^4q^2 + p^2q^4]}{10p^2q^2} = \frac{10p^2q^2[p^2 + q^2]}{10p^2q^2} = p^2 + q^2.$$

It is interesting to note that

$$p^2 + q^2 = p^2 + 2pq + q^2 - 2pq = (p + q)^2 - 2pq$$
$$= 1 - 2pq,$$

which is $\frac{1}{2}$ if $p = q = \frac{1}{2}$ and is bigger than $\frac{1}{2}$ otherwise.[2] So, if our independence assumptions are correct, World Series should end after 6 games at least as often as they go 7 games. The table on page 36 shows that this has not been the case for the past 100 years. There must be other factors at play.

I now verify the entries in the table on page 95.

The sequence of outcomes leading to the favored team winning the series in 4 games consists of Bernoulli trials with $n = 4$ and all successes. Therefore the probability is $_4C_4 \times p^4 q^0 = p^4$. This is the first entry in the second row of the table.

If the favored team wins the series in 5 games, then it must win 3 out of the first 4 games—and then win the 5th game. The first 4 games form Bernoulli trials with $n = 4$ and 3 successes, so the probability is $_4C_3 \times p^3 q$. Thus the probability of this occurring *and* the favored team winning the 5th game is the product

$$(_4C_3 \times p^3 q) \times p = 4 \times p^4 q.$$

This is the second entry in the second row of the table.

If the favored team wins the series in 6 games, then it must win 3 of the first 5 games—and then win the 6th game. This time the first 5 games form Bernoulli trials with

---

[2]This example is based on an observation in a paper by Groeneveld and Meeden, "Seven Game Series in Sports," *Mathematics Magazine*, 1975, pp. 187–192.

$n = 5$ and 3 successes, so the probability is ${}_5C_3 \times p^3q^2$. Thus the probability of the favored team winning the series in 6 games is

$$({}_5C_3 \times p^3q^2) \times p = 10p^4q^2.$$

If the favored team wins the series in 7 games, then it must win 3 of the first 6 games, Bernoulli trials with $n = 6$ and 3 successes, and then win the 7th game. The probability that this happens is

$$({}_6C_3 \times p^3q^3) \times p = 20p^4q^3.$$

The calculations for the probabilities of the favored team to lose the series in 4, 5, 6, and 7 games are similar. In fact, it just amounts to interchanging all the ps and qs.

The calculations for the table on page 97 are similar to those just given. For example, given that the favored team wins the first game, the conditional probability that it wins the series is the probability that it wins 3 games in a series of 6 possible games. Ambitious readers are invited to verify the entries of the table.

When Should You Stop Betting?
Double-or-Nothing

Money won is twice as sweet as money earned.

PAUL NEWMAN IN "THE COLOR OF MONEY"

Now we are ready to consider general questions about betting strategies like: "How likely is it that I can double my fortune before going broke?"

Consider again a game you'd like to repeat over and over, where the probability of winning is p each time and where the payoff is one dollar for every dollar bet. My favorite example is betting red in roulette where your odds are 9:10 and $p \approx 0.4737$. This is almost a sequence of Bernoulli trials, because the outcomes of the trials are independent, except that I'm not specifying the number n of games to be played. If I specified n, then that would specify exactly how long you may play. But now I want you to double your investment—or lose it all trying. Notice how *I* make the rules, but then expect *you* to do the playing. That way I keep my money.

Here I assume that you will begin with a certain amount of money, which I will call your "fortune," and agree to play until you *double* your fortune or *lose it all*. The question is: How likely is it that you'll double your fortune, rather than losing it all? This is an old problem and is called the "ruin problem" because it is assumed that you are ruined if you lose your fortune. In fact, one of Bernoulli's brothers, Niclaus or Nikolaus, contributed to this problem in the early 1700s.

You may be acquainted with the conventional wisdom that if you want to maximize your chances of success with a double-or-nothing strategy, you should just bet the whole fortune in one bet. This assumes that your odds *b:a* are unfavorable, as in a casino; i.e., *b* is less than *a* and so *p* is less than $\frac{1}{2}$. Of course, betting your whole fortune in one bet is no fun and you may as well stay home and avoid the risk.

In our roulette example, where you bet on red, your odds are 9:10 so that $b = 9$, $a = 10$, and $p \approx 0.4737$. Suppose that you have \$640 that you would like to double using the double-or-nothing strategy. Then you'll play until your fortune has grown to \$1280 or until you've lost it all.

Should you bet \$640 all at once? Or \$320 per bet? Or \$64 per bet? Or just \$1 per bet (if they'll let you bet such a low amount)?

First, let's clarify the question. If you bet \$640 all at once, you are trying to double-or-nothing in 1 step, and this is mathematically boring because it is obvious that your odds for winning are *b:a* (i.e., 9:10) and your probability of success is $p \approx 0.4737$.

If you bet \$320 at each bet, you are trying to double-or-nothing by getting 2 steps ahead or behind, in the sense that you won't quit until your number of wins is either 2 *more* than your number of losses, in which case you're happy, or your number of wins is 2 *less* than your number of losses, in which case your "fortune" is gone. Similarly, if you bet $64 per bet, you won't quit until your number of wins is either 10 more or 10 less than your number of losses (because $10 \times \$64 = \$640$). Likewise, if you could bet $1 per bet, you'd need your number of wins to be either 640 ahead or behind your number of losses before you've doubled your fortune or lost it all. This might take a long time.

In all these cases, the odds $b{:}a$ and probability $p = \frac{b}{b + a}$ for success are the same. The only difference in the strategies is the number s of steps ahead or behind needed to either double your fortune or lose it all. In the four examples where we bet $640, $320, $64, or $1 per bet, the values of s are 1, 2, 10, and 640. All that matters is your odds $b{:}a$ for success on each play and the number s of steps you need to be ahead or behind before quitting. The dollar values discussed are irrelevant.

If the conventional wisdom is correct, your chances of doubling your fortune are best when s is smallest.

The answers to the questions posed earlier will follow from the next result.

Double-or-Nothing Story

If the odds are $b{:}a$ for winning each time, then the odds for doubling your fortune before losing it, where you need your number of wins to be s larger than your number of losses, are $b^s{:}a^s$.

Note how this statement is straightforward and attractive, and maybe even intuitive when you think about it, because it is stated in terms of odds. Way back in Chapter 2, I suggested that probability, rather than odds, is almost always the better setting from the point of view of mathematics. This story is an *exception*. Of course, I could restate the story in terms of probability, instead of odds, but then the result would be less attractive and intuitive. This was the first result about probability that I realized was more understandable using the language of odds instead of the language of probability. Moreover, the explanation is more understandable in terms of odds.

To repeat myself, the odds are 9:10 of winning each game of roulette if you bet on red. These are the odds of doubling your fortune in *one* step. For $s = 1$, the Double-or-Nothing Story doesn't tell us anything we didn't already know.

The odds of doubling your fortune when you need to get 2 wins ahead (i.e. $s = 2$) are $9^2:10^2$ or 81:100. This is close to 8:10 odds, so these odds are worse than 9:10. To double your \$640 fortune, you are better off trying to do it in one step than betting \$320 on each bet. The conventional wisdom seems to be correct!

It only gets worse if you try to double your fortune when the required number s of wins ahead of your losses is bigger. If you bet \$64 on each bet and your fortune is \$640, then $s = 10$ and your odds are $9^{10}:10^{10}$, which is close to 1:3 because

$$\frac{9^{10}}{10^{10}} = (.9)^{10} \approx .35 \approx \frac{1}{3}.$$

Since $p = 0.4737$ is pretty close to $\frac{1}{2}$, it is not unreasonable to hope to double your fortune with $s = 10$, but the odds are about 3:1 against it.

What happens if you try to double your \$640 fortune by betting \$32 each time? Then $s = 20$ and the odds for doubling your fortune are $9^{20}:10^{20}$, which is close to 1:8. The odds are about 8:1 against success. Not great, but not hopeless either.

What about betting $10 a bet, so that $s = 64$? The odds $9^{64}:10^{64}$ are close to 1:848, which is pretty hopeless.

You may be dying to know how grim the situation is if you bet $1 per bet, so that $s = 640$. The odds $9^{640}:10^{640}$ involve astronomical numbers. The number 9^{640} is too big for my calculator, but it can handle $\left(\frac{10}{9}\right)^{640}$. The number $\left(\frac{10}{9}\right)^{640}$ is big; it's about

$$B = 192,661,065,959,926,746,989,310,923,556.$$

Therefore 10^{640} is about B times as big as 9^{640}. That's grim! This is way worse than any lottery. For all practical purposes, there is *no chance* of doubling your $640 fortune betting a dollar a game.

How much would it help if you found a game with the probability of success *really* close to $\frac{1}{2}$, say $p = 0.495$ so that the odds are 99:101? You can apply the Double-or-Nothing Story as well as I can. I'll just give some answers with $s = 10, s = 20, s = 64$, and $s = 640$. The corresponding odds are close to 4:5, 2:3, 2:7, and 1:362,372.

The Double-or-Nothing Story is also true if $b = a = \frac{1}{2}$, in which case the game is fair. No matter what s is, $b^s = a^s$ and so these odds are 1:1 odds. In words, if the original game is fair, then any double-or-nothing scheme also will be fair. This is intuitively correct.

The Double-or-Nothing Story is true even when the odds $b:a$ are favorable so that $p > \frac{1}{2}$. In this case, your odds of success get better and better as s increases. This is why casinos, which have all the time in the world, always make money in the long run.

Next I will explain why the Double-or-Nothing Story is true. You may wish to skip to the discussion of other stopping strategies on page 111.

When giving odds $b{:}a$, it's traditional in the real world to have b and a be whole numbers, and this is helpful in interpreting betting strategies. But there is no mathematical reason to require b and a to be whole numbers. In fact, it is convenient to observe the following. If the odds are $b{:}a$ and if, as usual, p is $\frac{b}{b+a}$ and q is $1 - p = \frac{a}{b+a}$, then

- The odds $b{:}a$ and $p{:}q$ are the same.

Moreover,

- For any s, the odds $b^s{:}a^s$ and $p^s{:}q^s$ are the same.

The second statement is true because the ratios

$$\frac{b^s}{a^s} \quad \text{and} \quad \frac{p^s}{q^s} = \frac{\left(\dfrac{b}{b+a}\right)^s}{\left(\dfrac{a}{b+a}\right)^s} = \frac{b^s}{a^s}$$

are equal. If you replace s with 1 in the second statement, you will get the first statement.

Because of these observations, I restate the Double-or-Nothing Story as

Double-or-Nothing Story II

If the odds are $p{:}q$ for winning each time, then the odds for doubling your fortune before losing it, where you need your number of wins to be s larger than your number of losses, are $p^s{:}q^s$.

This version of the story is easier to explain because p and q represent certain probabilities. I will illustrate all the ideas by considering the case $s = 3$ and then briefly the case $s = 4$.

Suppose that $s = 3$. I need to explain why the probability N of doubling the fortune in this case and the probability D of going broke in this case have the same ratio as p^3 to q^3. In other words, I need to convince you that

$$\frac{N}{D} = \frac{p^3}{q^3}.$$

I used the symbols N and D because N is the numerator and D is the denominator.

The numerator N is the sum of all the probabilities of all the ways we can end up doubling the fortune. Here's a typical way this can happen, where I'm writing S for success (winning) and F for failure (losing) for each game:

$$S\,F\,F\,S\,F\,S\,S\,F\,S\,S\,F\,S\,S.$$

The key fact about this sequence of successes and failures is that we were never behind or ahead 3 games, so we hadn't doubled our fortune or gone broke, until the 13th game. In this example, we fell behind 1 on the third and fifth games and we got 2 ahead on the 10th game. Of course we were ahead 2 on the 12th game too. Finally, on the 13th game we were 3 games *ahead* and had doubled our fortune. What's the probability of this exact sequence of outcomes, which leads to doubling our fortune? Because the events are independent, it is the following product:

$$p \times q \times q \times p \times q \times p \times p \times q \times p \times p$$
$$\times q \times p \times p = p^8 \times q^5. \qquad (*)$$

Here is the crux of the argument. To each such term in the sum for the numerator N, there is a natural "opposite" sequence of outcomes that give a term in the sum for the denominator D, namely

$$F S S F S F F S F F S F F$$

for our example, where each S turned into an F and each F turned into an S. With this sequence, we would never be ahead or behind more than 2 games until the 13th game, and then we would be 3 games *behind* and lose our fortune. The probability of this exact sequence of outcomes, which led to our ruin, is

$$q \times p \times p \times q \times p \times q \times q \times p \times q \times q$$
$$\times p \times q \times q = p^5 \times q^8. \qquad (**)$$

Now observe that the ratio of the number in (*) to this number in (**) is

$$\frac{p^8 \times q^5}{p^5 \times q^8} = \frac{p^3 \times p^5 \times q^5}{p^5 \times q^5 \times q^3} = \frac{p^3}{q^3}.$$

This is because the probability in (*) needed three extra ps and the probability in (**) needed three extra qs. This remark applies to every sequence of Ss and Fs, no matter how short or long, which gives a term in N and its "opposite," which gives a term in D. In other words, every term in the sum N is $\dfrac{p^3}{q^3}$ times its corresponding term in D. So the entire sum N is $\dfrac{p^3}{q^3}$ times the entire sum D. Therefore

$$N = \frac{p^3}{q^3} \times D \text{ and}$$

$$\frac{N}{D} = \frac{p^3}{q^3},$$

as I claimed.

Each of the sequences of games in the $s = 3$ case consists of an odd number of games. Our examples happened to involve 13 games. This is because you can't be exactly 3 games ahead or behind after an even number of games. This comment applies whenever s is odd. The only difference in the case that s is even, say $s = 4$, is that all sequences of games leading to doubling our fortune or going broke will have to consist of an even number of games. The argument for $s = 4$ would be almost identical to the argument for $s = 3$, and this time we'd end up with

$$\frac{N}{D} = \frac{p^4}{q^4}.$$

Infinite Sums

In the discussion of the numerator N and denominator D of the ratio of probabilities of "double-or-nothing," I worked with sums of certain probabilities. There were an infinite number of terms because we have no control over how long it will take to either double our fortune or lose it all. It could take a very long time, though this is very unlikely. But because there's no limit on the number of tries it could take, we cannot ignore any of the terms.

The good news is that most of the terms in the infinite sums are quite small and the *infinite sums actually represent numbers*. This all makes sense because of the theory of limits and infinite series, a topic usually first encountered in calculus. When I took calculus, a long time ago, this theory was the first thing in calculus that really excited me. Go figure!

Note that, in explaining the Double-or-Nothing Story, I did not need to calculate the numerator N and denominator D. All I needed to know was their ratio

$$\frac{N}{D} = \frac{p^s}{q^s}.$$

Because N represents the probability of doubling ones fortune and D represents the probability of going broke, it would be nice to know their values. As I will argue later, the sum $N + D$ is 1. So I can solve for N and D from the equations

$$\frac{N}{D} = \frac{p^s}{q^s} \quad \text{and} \quad N + D = 1.$$

First I solve for N in the first equation and plug the result in the second equation, to get

$$\frac{p^s}{q^s} \times D + D = 1, \quad \left[\frac{p^s}{q^s} + 1\right] \times D = 1,$$

$$\text{and so} \quad \frac{p^s + q^s}{q^s} \times D = 1;$$

hence

$$D = \frac{q^s}{p^s + q^s}.$$

Then

$$N = 1 - D = 1 - \frac{q^s}{p^s + q^s} = \frac{[p^s + q^s] - q^s}{p^s + q^s} = \frac{p^s}{p^s + q^s}.$$

Now that we see the formulas for N and D, it is clear that $N + D = 1$ and that the ratio $\dfrac{N}{D}$ is exactly $\dfrac{p^s}{q^s}$.

Here is why $N + D = 1$. It is imaginable that the process would go on forever, because one could conceivably alternately win and lose forever. However, it can be shown that the probability that the process will go on forever is 0. This is beyond the scope of this book, but when there are an infinite number of possibilities, there can be possibilities that have probability 0. Anyway, because the probability that the process will go on forever is 0, the probability that one will either double ones fortune *or* go broke is 1. In symbols, $N + D = 1$.

Other Stopping Strategies

Again, consider a game you'd like to repeat over and over, where the probability of winning is p each time and where the payoff is one dollar for every dollar bet. Again, assume that your strategy will be to quit and keep your fortune if you get s dollars ahead. Unlike the preceding discussion, though, let's not assume that you begin with s dollars with a "double-or-nothing" strategy. Instead, let's assume that your fortune, that you are willing to risk, is t dollars. In other words, if you are betting $1 per game, your strategy is to stop when you are $s ahead or $t behind. I will call this strategy a **stopping strategy**, though probabilists call this a "stopping time."

Note that the stopping strategy where $s = t$ is exactly the double-or-nothing strategy. I will reconcile the stories by the end of this chapter.

The fair case when $b = a$ and $p = \frac{1}{2}$ is quite elegant.

Fair Case

If $p = \frac{1}{2}$ and your strategy is to quit the first time you are s wins ahead or t wins behind, then $\frac{t}{s+t}$ is the probability of success, i.e., stopping s wins ahead. The probability of failure is then $\frac{s}{s+t}$, so the odds for success are $t{:}s$.

The Fair Case is pretty obvious if $s = t$, as we noted on page 105.

The simplest nonobvious version of the Fair Case is when $s = 1$ and $t = 2$. Then the probability of getting 1 win ahead before getting 2 wins behind is

$$\frac{t}{s+t} = \frac{2}{1+2} = \frac{2}{3}.$$

Of course the probability of getting 2 wins behind before getting 1 win ahead is $\frac{1}{3}$. In other words, the odds are 2:1 that you'll get 1 win ahead before getting 2 wins behind. It is twice as likely that you'll end up 1 ahead as it is that you'll end up 2 behind.

A similar remark applies whenever t is twice as big as s, i.e., whenever $t = 2 \times s$, because in this case

$$\frac{t}{s+t} = \frac{2 \times s}{s + 2 \times s} = \frac{2 \times s}{3 \times s} = \frac{2}{3}.$$

For example, the odds are 2:1 that you'll get 5 wins ahead before getting 10 wins behind.

These strategies are all fair. For example, if $s = 5$ and $t = 10$, then the expectation is

$$\mathbf{Pr}(\text{winning}) \times 5 - \mathbf{Pr}(\text{losing}) \times 10$$
$$= \frac{2}{3} \times 5 - \frac{1}{3} \times 10 = 0,$$

which tells us that the strategy is fair.

To actually verify even the simplest nonobvious Fair Case is something of a chore. Let's just admire the simplicity of its statement. For the general case, where p is not equal to q, the formula is a lot more complicated. As before, I'll write q for the probability of losing each time. So $q = 1 - p$.

General Case

If p is not equal to q and your strategy is to quit the first time you are s wins ahead *or* t losses behind, then

$$\frac{\left(\frac{p}{q}\right)^s - \left(\frac{p}{q}\right)^{s+t}}{1 - \left(\frac{p}{q}\right)^{s+t}}$$

is the probability of success, i.e., stopping s wins ahead.

Yet again we return to the roulette example, where $p = \frac{18}{38}$ and $q = \frac{20}{38}$. In this case

$$\frac{p}{q} = \frac{18/38}{20/38} = \frac{18}{20} = \frac{9}{10} = 0.9.$$

Suppose that a roulette player wants to win $100 and is willing to risk losing $700 trying. He plans to bet $100 on each game. This is the case $s = 1$ and $t = 7$, where $\frac{p}{q}$ is still 0.9. Then the probability of success is

$$\frac{(0.9)^1 - (0.9)^8}{1 - (0.9)^8} \approx \frac{0.9 - 0.430}{1 - 0.430} = \frac{0.470}{0.570} \approx 0.825.$$

The player's chance of success is over 80 percent and his odds for success are close to 5:1, since $\frac{5}{6} \approx 0.833$. The problem is that, when he loses, he will lose seven times what he wins, when he wins. Following this strategy, his expectation is

$$\mathbf{Pr}(\text{winning}) \times 100 - \mathbf{Pr}(\text{losing}) \times 700$$

$$\approx \frac{5}{6} \times 100 - \frac{1}{6} \times 700 = -\frac{200}{6} \approx -\$33.$$

There are new wrinkles with this stopping strategy, but the expectation is still negative.

Let's suppose that the roulette player is more greedy. He still plans to bet $100 per game and he wants to win $500. He is willing to risk $1000. This is the case $s = 5$ and $t = 10$. The probability of success is

$$\frac{(0.9)^5 - (0.9)^{15}}{1 - (0.9)^{15}} \approx \frac{0.590 - 0.206}{1 - 0.206} = \frac{0.384}{0.794} \approx 0.484.$$

Thus the odds are nearly 1:1 that the player will be successful. This isn't so great, though, because each time he loses, he loses twice as much as he wins, when he wins. His expectation is now

$$\mathbf{Pr}(\text{winning}) \times 500 - \mathbf{Pr}(\text{losing}) \times 1000$$
$$\approx \frac{1}{2} \times 500 - \frac{1}{2} \times 1000 = -\$250.$$

This is not very encouraging. He would be better off betting \$500 each time. This would be the case $s = 1$ and $t = 2$. The probability of success would be

$$\frac{(0.9)^1 - (0.9)^3}{1 - (0.9)^3} \approx \frac{0.9 - 0.729}{1 - 0.729} = \frac{0.171}{0.271} \approx 0.631,$$

which isn't too far from $\frac{2}{3}$, the probability of success if $p = \frac{1}{2}$. His expectation would be

$$\mathbf{Pr}(\text{winning}) \times 500 - \mathbf{Pr}(\text{losing}) \times 1000$$
$$\approx 0.631 \times 500 - 0.369 \times 1000 = -\$53.50.$$

This expectation is not as bad as betting \$100 per game.

Let's now suppose that the roulette player is even more greedy. He wants to win \$700 by playing \$100 per game, and he is only willing to risk \$700. This is the case $s = 7$ and $t = 7$. Then the probability of success is

$$\frac{(0.9)^7 - (0.9)^{14}}{1 - (0.9)^{14}} \approx \frac{0.250}{0.771} \approx 0.324 \approx \frac{1}{3}.$$

With this strategy, the expectation is approximately

$$\frac{1}{3} \times 700 - \frac{2}{3} \times 700 = -\frac{700}{3} \approx -\$233.$$

Again the expectation is negative.

Actually, this last strategy, where $s = t$, is exactly the Double-or-Nothing strategy. The odds for roulette are 9:10, so $b = 9$ and $a = 10$. By the Double-or-Nothing Story on page 103, the odds for winning are $9^7 : 10^7$. Since $9^7 = 4,782,969$ and $10^7 = 10,000,000$, the odds are about 48:100 or 12:25. So by the rule on page 23, the probability of success is about

$$\frac{12}{12 + 25} = \frac{12}{37} \approx 0.324.$$

This agrees with our previously mentioned calculation. Later in this chapter I will elaborate on why the two results always give the same answer in the $s = t$ Double-or-Nothing case.

In the last two examples, each stopping strategy had a negative expectation. I am not surprised because it is a major result in probability theory that

There is no stopping strategy that can change a losing situation $\left(p < \frac{1}{2}\right)$ into a winning situation.

This important result was first recognized by Richard von Mises (1883–1953), who was an outstanding Austrian applied mathematician and probabilist. A fancier version of this result assures us that there is no winning strategy to roulette (or in other situations where the assumptions in this chapter hold) if the strategy at each step is based entirely on information already at hand. The assumptions don't allow for clairvoyance or extraneous information such as knowing about faulty equipment or dishonest games.

Now let's assume that $p < \frac{1}{2}$ (the bad news), but that you are very rich (the good news), and that you are willing to risk a very large amount t. Since $\frac{p}{q}$ is less than 1, $\left(\frac{p}{q}\right)^t$ and $\left(\frac{p}{q}\right)^{s+t}$ will be very small because they are high powers of $\left(\frac{p}{q}\right)$. Therefore the probability of success is

$$\frac{\left(\frac{p}{q}\right)^s - \left(\frac{p}{q}\right)^{s+t}}{1 - \left(\frac{p}{q}\right)^{s+t}} = \frac{\left(\frac{p}{q}\right)^s\left[1 - \left(\frac{p}{q}\right)^t\right]}{1 - \left(\frac{p}{q}\right)^{s+t}} \approx \frac{\left(\frac{p}{q}\right)^s[1 - 0]}{1 - 0} = \left(\frac{p}{q}\right)^s.$$

This nice approximation will appear in our next result.

Now let's move into the realm of fantasy and assume that you are infinitely rich. Even Bill Gates is not infinitely rich, though he's close. Let's also assume that p is less than $\frac{1}{2}$, so that $\frac{p}{q}$ is less than 1. Then the next claim makes sense (well, except for assuming that you're infinitely rich).

Infinitely Rich Case

If $p < \frac{1}{2}$, you are infinitely rich and your strategy is to quit when you are s wins ahead, then

$$\left(\frac{p}{q}\right)^s$$

is the probability that you'll be successful.

This means that, with this probability, you will eventually be s wins ahead and quit. The alternative is that you will never get s wins ahead and will play forever. Moreover, it's a probabilistic certainty that your losses will get larger and larger. Of course, since you are infinitely rich, this won't bother you much. Note that $\left(\frac{p}{q}\right)^s$ gets smaller and smaller as s increases. This makes sense because, for example, it's harder to get 5 wins ahead than it is to get 2 wins ahead.

We return to the roulette example, so that $\frac{p}{q} = 0.9$. Suppose that you only want to get one win ahead, i.e., $s = 1$. Then the probability is $(0.9)^1 = 0.9$ that you will be successful. I repeat: There's a 90 percent chance that you will eventually succeed. Moreover, if this happens it will probably happen fairly quickly. Indeed, you might get ahead on the very first try. You'll probably succeed, and impress your friends, if you bet $1000 on each game of roulette and quit the first time you get ahead. The trouble with this strategy is what happens the "other 10 percent of the time" when you will play forever and lose more and more. "Other 10 percent of the time" doesn't make a lot of sense here. It's somewhat easier to visualize *ten* infinitely rich gamblers. We're not sure, of course, but we would expect nine of them to win their $1000. The other one would lose more and more forever!

As another example, let's assume that you are more greedy and won't quit playing roulette until you are seven wins ahead. Then $s = 7$ and the probability of success is $(0.9)^7 \approx 0.478$. This time there's a little less than 50 percent chance of success. If you're betting $1000 on each game, there's about a 48 percent chance that you'll eventually get $7000 ahead. That's nice, but there's a 52 percent chance that you'll lose forever!

Time for some explanations. Here is the General Case again.

General Case

If p is not equal to q and your strategy is to quit the first time you are s wins ahead *or* t losses behind, then

$$\frac{\left(\frac{p}{q}\right)^s - \left(\frac{p}{q}\right)^{s+t}}{1 - \left(\frac{p}{q}\right)^{s+t}}$$

is equal to the probability of success, i.e., stopping s wins ahead.

This result is a version of formula (2.4) in section XIV.2 of Feller's classic [16], with different notation. (The fraction is Feller's $1 - q_z$, where $z = t$ and $a = s + t$.) I won't explain this complicated formula, but I will show that it includes familiar cases.

As illustrated in the roulette example on page 115, if $s = t$, then this formula gives the same answers as the Double-or-Nothing Story. Here is why. With $s = t$, the probability of success is

$$\frac{\left(\frac{p}{q}\right)^s - \left(\frac{p}{q}\right)^{2s}}{1 - \left(\frac{p}{q}\right)^{2s}}.$$

This has the form

$$\frac{x - x^2}{1 - x^2} \quad \text{where} \quad x = \left(\frac{p}{q}\right)^s.$$

Since in general

$$\frac{x - x^2}{1 - x^2} = \frac{x(1 - x)}{(1 - x)(1 + x)} = \frac{x}{1 + x},$$

the probability of success simplifies to

$$\frac{\left(\frac{p}{q}\right)^s}{1 + \left(\frac{p}{q}\right)^s} = \frac{p^s}{q^s \times \left[1 + \left(\frac{p}{q}\right)^s\right]} = \frac{p^s}{q^s + p^s}.$$

This is exactly the expression we obtained for this probability on page 111, where it was denoted N.

I've already explained the $s = t$ case, and I now give a relatively simple explanation of the Infinitely Rich Case without using the General Case.

Infinitely Rich Case

If you are infinitely rich and your strategy is to quit when you are s wins ahead, then

$$\left(\frac{p}{q}\right)^s$$

is the probability that you'll be successful.

I first verify this for the case $s = 1$. Because I want, but don't know, the probability of ever getting ahead, I will temporarily denote this probability by x. In this paragraph, I will show that x satisfies the equation

$$x = p + qx^2.$$

If you ever get ahead, you will get ahead at the very first step (and this will happen with probability p) or you will lose at the very first step, then you will get even sometime, and then you will get ahead sometime later. This second possibility is a sequence of three independent events: "losing at the first step," "getting even again," and "getting ahead." The probability of "losing at the first step" is q. The probability of "getting even again" given that you lost at the first step is the same as the probability of "getting ahead" starting from the beginning (think of the sequence of games as starting at the second game instead of the first game). This probability is then the unknown x, and this is also the probability of "getting ahead" given that you are even again. Therefore the probability of the second possibility is the product of the probabilities of these three independent events, namely $q \times x \times x = qx^2$. Finally, the probability x of ever getting ahead is the sum $p + qx^2$.

Since $x = p + qx^2$, I need to solve the "quadratic equation"

$$qx^2 - x + p = 0.$$

I don't need the famous but unloved quadratic formula. Actually, I first used the quadratic formula, but when I saw the simple solutions, I realized that I could solve the equation by factoring, as follows. Since $1 = q + p$, we can write

$$qx^2 - x + p = qx^2 - (q + p)x + p$$
$$= qx^2 - qx - px + p = (qx - p)(x - 1).$$

So the solutions of $qx^2 - x + p = 0$ are the solutions of $(qx - p)(x - 1) = 0$. Either $qx - p = 0$, in which case $x = \frac{p}{q}$, or else $x - 1 = 0$, in which case $x = 1$. It is believable, and can be verified, that there's no guarantee that you'll ever be ahead. So x is not equal to 1, and the probability x of ever getting 1 win ahead is $\frac{p}{q}$. Since $\frac{p}{q} = \left(\frac{p}{q}\right)^1$, this explains the Infinitely Rich Case for the case $s = 1$.

Now consider the case $s > 1$. The probability of ever getting s steps ahead is the probability of getting one step ahead, and then getting two steps ahead, and then getting three steps ahead, and so forth, until we're s steps ahead. These s events are independent and each of them has probability p, because one can think of the sequence as starting over after achieving each level. Thus the probability of ever getting s steps ahead is the product of s copies of p, i.e., p^s. So the Infinitely Rich Case holds as stated.

If the house has an advantage, we've seen that the best double-or-nothing strategy is to bet all of one's fortune and see what happens. Even when we looked at more general stopping strategies, we still didn't find a way to get the advantage. The bottom line is that there's no preplanned strategy that can make a losing situation into a winning situation.

What About Streaks?
Statistics

Statistics are used like a drunk uses a lamppost—for support, not illumination.

VIN SCULLY, DODGERS' SPORTS ANNOUNCER

I'll bet 4:1 that you thought this whole book was already about statistics. So why is this late chapter suggesting otherwise? In this book, when we haven't been looking at probability, we've been looking at descriptive statistics, statistics like *AVG* that summarize data. Occasionally I discussed informally what one might infer from them. The latter was my informal way of touching on inferential statistics. The tools of inferential statistics are used to clarify and quantify information based on descriptive statistics like *OBP, SLG,* and *OPS.* First, I discuss the word "statistics."

státis'tics, *n.pl.* 1. facts or data of a numerical kind, assembled, classified, and tabulated so as to present significant information about a given subject. 2. [*construed as sing.*] the science of assembling, classifying, and tabulating such facts or data. WEBSTER'S NEW TWENTIETH CENTURY DICTIONARY, 2nd edition

As the definition shows, "statistics" is both singular and plural, depending on the usage. The old book [29, *Chance, Luck and Statistics,* page 5] has a nice explanation:

The word "statistics" is unusual. As a plural noun it means collections of sets of facts that are related, such as wheat yields per acre or the number of births per year. In practice, it is customary to restrict its meaning to facts that are numerical, or can in some way be related to numbers. As a singular, collective noun statistics means the science of collecting or selecting statistical facts, sorting and classifying them, and drawing from them whatever conclusions may lie buried among them. The facts, for example, might have to do with the incidence of a certain disease. The conclusion might be that the disease thrives only when sanitation is poor. The singular form "statistic" is sometimes used to mean a single statistical fact. To avoid confusion it is important to keep these distinctions in mind.

The concepts in Chapter 1 provide statistics in the plural sense. There are massive helpful books full of all sorts of such statistics; see Bill James' [24, *All-Time Baseball Sourcebook*], John Thorn and Pete Palmer's [42, *The Hidden Game of Baseball*] and [43, *Total Baseball*], as well as [23, *Major League Handbook*]. Inferential statistics, which I will touch on in this chapter, are part of statistics in the singular sense.

Before proceeding, I want to mention the book [40, *Baseball by the Numbers*]. The subtitle of this book is "How Statistics Are Collected, What They Mean, and How They Reveal the Game." This book is written for the "statistically knowledgable" baseball fan, but "neophytes will definitely benefit from reading" the first few chapters. I recommend this book to baseball fans who want to do statistical research in baseball, but who are not professional statisticians. For fans who know no statistics at all, beyond what they've gleaned from this book, I recommend my favorite freshman statistics text [33, *The Basic Practice of Statistics*]. I like this book because it focuses on the ideas of statistics rather than on the calculations, which any decent spreadsheet program can do. The author is very frank about the good and bad practices of statistics.

In general, inferential statistics is a sophisticated field that clarifies what can be inferred or predicted, to what accuracy, and with what confidence. Inferential statistics is a science with clear if not-always-stated assumptions. This is quite different from what we might call "qualitative judgments." For example, Ted Williams' lifetime *AVG* and *SLG* are .344 and .634, while Joe DiMaggio's numbers are .325 and .579. In fact, Ted Williams lifetime *SLG* of .634 is second only to Babe Ruth; the Babe's lifetime *SLG* was .690. These are facts. Do these facts prove that Williams was a better hitter than DiMaggio? Probably, even though these aren't the only aspects of hitting that are important. In any case, the conclusion is at best a qualitative judgment. Do these facts prove that Williams was a better ballplayer than DiMaggio? Not at all. Even though hitting is very important, there's a lot more to being a great ballplayer than hitting.

The first topic that I'll cover in this chapter is correlation. In ordinary conversation, "correlation" usually refers to a mutual relationship or connection between two things. The meaning in statistics is not that different. The two "things" will be two sets of numerical data that are linked or paired in some natural way. For example, we'd expect that the heights and weights of ten-year-old boys in the United States would be somewhat correlated in the sense that, generally speaking, taller boys would tend to weigh more and, vice-versa, heavier boys would tend to be taller. We might check this using fourth graders at some school. To visualize the correlation, we might graph the pairs of numbers (height, weight), where there would be one pair of numbers for each ten-year-old boy. Linked pairs of numbers are called *paired data*.

As another example, we would hope that baseball teams' wins-per-season would correlate with their player payrolls. That is, we would expect that, generally speaking, the teams with higher payrolls would have better seasons. Here the paired data consist of pairs of numbers (wins, payroll), with a pair for each team. As with all interesting correlations, there will be exceptions to the rule. An uninteresting correlation exists between Fahrenheit and Celsius temperatures collected at some particular times or places. This is uninteresting, because we have a formula connecting them, so we know their relationship exactly.

We would also like to think that pitchers' salaries and earned-run-averages (*ERA*s) are correlated, but *in reverse*. That is, good pitchers should get high salaries and have *low ERA*s, while pitchers with high *ERA*s should generally be paid less. Such reverse correlations will be called "negative correlations." We would not want to say there is "no correlation" because there is (or should be) a connection; it just goes in the opposite direction from the more familiar correlations.

You should not be surprised that, given some paired data of interest, statisticians assign a number that measures how strong the correlation is. This will be a number r between −1 and 1, which is called the *correlation*. It turns out that the formula for r is a big mess, and it is hard to understand its significance from its formula. Instead, I will state the key properties of correlation and give several examples.

Properties of Correlation

Given paired data, the correlation is a number r between −1 and 1. If r is positive, there is a correlation. If r is very close to 1, then the correlation is strong. If r is close to 0, the correlation is weak. If r is negative, we have so-called negative correlation, as in our example involving pitchers' *ERA*s and their salaries.

In words, correlation measures the strength and direction of the linear relationship of the paired data.

People talk about strong and weak correlations, but authors seem reluctant to clarify what they mean. I'm fearless, or foolish, so here are my rough usages of these words:

strong correlation: $0.70 \leq r \leq 1.00$
weak correlation: $0.30 \leq r \leq 0.70$
very little correlation: $-0.30 \leq r \leq 0.30$
weak negative correlation: $-0.70 \leq r \leq -0.30$
strong negative correlation: $-1.00 \leq r \leq -0.70$

As I mentioned, I'm confident there's a positive correlation between ten-year-old boys' heights and weights. But I would need to collect real data (using real boys), and calculate r, to know whether the correlation is strong or weak, or even very weak.

Even casual experience with box scores reveals that there is a positive correlation between the number of hits and the number of runs obtained by each team in a game. In general, the more hits a team gets, the more runs it will get, and vice-versa. This isn't very exciting, but I checked three sets of data and found a fairly strong correlation. For a typical day in 2000 (May 10), I found the correlation to be ≈ 0.859. For a typical day in 2003 (August 15), the correlation was ≈ 0.774; 30 teams played 15 games that day. Postseason games are somewhat different, so I checked the correlation for all 38 postseason games in 2003 and found it to be ≈ 0.707. I used a spreadsheet for these and similar calculations of correlation.

Most offensive statistics for hitters will have *positive* correlations, because they all measure successful at-bats, though some might not, for example, triples, where speed is important, and home runs, where raw power is important. In fact, for the top 50 Major League hitters of 2003, the correlation of triples and home runs was slightly negative, ≈ -0.147. By the way, there was a weak correlation, 0.519, between strike-outs and home runs.

Also, we would expect starting pitchers' number of wins and salaries to have a positive correlation. See page 131, where I provide some imaginary data.

We would expect pitchers' winning percentages and their earned-run-averages, *ERAs*, to have a *negative* correlation, because the better pitchers should have higher winning percentages and lower *ERAs*. For the top 50 pitchers in 2003, based on their *ERAs*, the correlation was negative, as expected, namely ≈ -0.347, but this is a surprisingly weak correlation. The correlation of *ERAs* with total wins was even weaker, an almost insignificant ≈ -0.160. I am very surprised.

As I observed on page 11, many experts feel that $BRA = OBP \times SLG$ is a better measure of offensive effectiveness than $OPS = OBP + SLG$. They feel this way because BRA correlates better with statistics that measure runs created. It boils down to the fact that BRA gives more credit to a high OBP, other things equal. Here's why. In all cases that I know of, SLG is higher than OBP, though sometimes they are close. Thus, if two players have equal OPS, then the one with a higher OBP will have his OBP closer to his SLG than the other player, and this will make his BRA higher than the other player's. For example, suppose that player A has $OBP = .250$ and $SLG = .475$, while player B has $OBP = .350$ and $SLG = .360$. Then player A's $OPS = .725$ is better than player B's $OPS = .710$, yet player B's $BRA = .126$ is better than player A's $BRA \approx .119$. Player B's higher OBP pushes his BRA above that of player A. I agree with this analysis, but I don't think the use of OPS will give much different results because BRA and OPS are generally strongly correlated.

The correlation of OPS and BRA for the top 50 Major League hitters of 2003 is $\approx .992$, which is very close to 1. The ordering of the players is about the same whether one uses OPS or BRA. By any measure, the top three hitters were Barry Bonds, Albert Pujols, and Todd Helton, who were all in the National League. Here are the key statistics for these three players:

Top Three Hitters in 2003

	OBP	SLG	OPS	BRA	AVG
Barry Bonds	.529	.749	1.278	.396	.341
Albert Pujols	.439	.667	1.106	.293	.359
Todd Helton	.458	.630	1.088	.289	.358

These were the only players with *OPS* above 1.019 or *BRA* above .253.

The correlation of *OPS* and *BRA* for the fourteen 2003 Boston Red Sox players who had at least 100 official at-bats was ≈.996. I also noticed that the ordering of the fourteen players based on *OPS* and *BRA* interchanged only two players, as follows.

Johnny Damon and Doug Mirabelli

	OBP	SLG	OPS	BRA
Johnny Damon	.345	.405	.750	.1397
Doug Mirabelli	.307	.448	.755	.1375

Because *BRA* gives more credit to on-base-percentage than *OPS*, Damon is the better offensive player according to *BRA*, even though *OPS* gives the nod to Mirabelli. But, in fact, both of the numbers are close for both of the players.

A friend suggested that the *OPS* and *BRA* might correlate less in the 1930s, say, when players didn't work as hard to get on base via walks. I picked the year 1937 and calculated the correlation for these statistics for the 98 players who had at least 400 at-bats. It was ≈.996, which is the same as what we found for the 2003 Red Sox players.

The table below gives correlations of some 2003 Major League offensive *team statistics* with their winning percentages. Note that team *OBP*s actually correlated slightly better with team winning percentages than the preferred

Correlations of 2003 Team Statistics

	HR	OBP	SLG	OPS	BRA	AVG
With teams' winning percentages	0.387	0.655	0.578	0.625	0.628	0.554

measures, team *OPS*s and team *BRA*s. Team home run totals had the weakest correlation with team winning percentages. Also, the correlation between team *OPS* and team *BRA* was ≈0.9992, which is very high indeed.

When I introduced the idea of correlation, I used the word linear to describe the relation of the paired data. This means that the correlation r measures how closely the graphed data fit a straight *line*. For example, Fahrenheit and Celsius temperatures are perfectly correlated with $r = 1$. This is because, if we graph the data, all the points will lie on a straight line. In fact, $F = \frac{9}{5} \times C + 32$ is an equation of the straight line that all the points will lie on.

If $r = 1$ for some paired data, and the paired data are graphed, the points on the graph will lie on a straight line that is rising from left to right. If $r = -1$, the points on the graph will lie on a straight line that is falling from left to right.

Let's pretend that, under a new Commissioner of Baseball, each year all starting pitchers are paid as follows. In the American League (AL), these pitchers are paid $500,000 plus $300,000 per win. In the National League (NL), they are paid $500,000 plus $200,000 per win, plus a bonus of $1,000,000 for winning more than ten games.

In both leagues, salaries and wins would be correlated in the sense that the players with more wins would earn more money. In the AL, the correlation would be perfect, i.e., $r = 1$, because all the data points would lie on a straight line. In fact, the equation of the straight line is

$$s = 500,000 + 300,000 \times w,$$

where w represents the number of wins, and s represents the salary.

However, the data points for the NL would not lie on a straight line. For pitchers with at most ten wins, their data points would lie on the straight line

$$s = 500,000 + 200,000 \times w,$$

but for the other pitchers, the data points would lie on the line

$$s = 1,500,000 + 200,000 \times w,$$

because of the million dollar bonus. Moreover, the correlation r would not equal 1, though the actual value of r would depend on how many wins each pitcher had. To greatly simplify our story from fairyland, suppose there were 100 starting pitchers, that 50 of them won exactly 6 games, 25 of them won exactly 12 games, and the other 25 won exactly 18 games. Then the correlation r turns out to be $\approx.989$.

When there's a strong correlation for paired data, there is a straight line that best approximates the data.[1] It is called the (least) *regression line*. This line can help us understand the data.

Using the 2003 Major League teams again, the correlation of the teams' runs and team wins is $\approx.604$, and the regression line is

$$w = 0.090 \times x + 12.2,$$

where x is the number of runs scored during the year and w is the number of team wins. Since 0.090×11 is close to 1, roughly speaking, it took about 11 runs to create a win.

---

[1]Actually, one can always find such a line, but it is most useful if the correlation is strong.

This is a pretty rough statement, though, because the correlation is not very strong. If we thought these data from 2003 were typical, we would say that it takes about 11 runs to create a win. In fact, some authors have remarked that it takes about 10 runs to create a win.

Sometimes the regression line is used to predict values of pairs of data not in the original study. If I'd omitted one Major League team in my calculations and it had scored 800 runs, using the regression line $w = 0.090 \times x + 12.2$, I would predict that the team would have won $0.090 \times 800 + 12.2 \approx 84$ games.

The next important concept I'll discuss is that of p-value. Before returning to baseball examples, it is convenient for me to return to the simple casino game of roulette. Let's suppose that a rich dowager is making large bets on red, and we suspect that the house is cheating her. Let's first suppose that she plays 10 games. For each game the probability of winning is about 0.474, so these games are Bernoulli trials with $p = 0.474$ and $n = 10$.

As noted on page 86, the expected number of wins is $10 \times 0.474 = 4.74$, but this can't mean that we expect *exactly* 4.74 wins. This means that if we repeated many many times the experiment of betting on red for 10 games, the *average* of the number of wins observed would be close to 4.74. We certainly wouldn't be surprised if she won 5 games, and I wouldn't be surprised if she won between about 3 and 7 games. But, suppose she didn't win any games? Or only 1 game? Or 2? Should we be surprised and suspicious that she's been cheated?

One way to mathematicize questions about surprise is to rephrase the questions and ask: How likely is the observation, given what we know? A useful concept is p-value. The intuitive idea, which we'll see is not quite correct, is that this is the probability of such an observation. With this in mind, let's focus on a single experiment where we observe our dowager making 10 bets on red. From the Key Results on page 86, we have

$$\mathbf{Pr}(\text{exactly } k \text{ wins in 10 games})$$
$$= {}_{10}C_k \times (.474)^k \times (.526)^{10-k}$$

for k between 0 and 10. The following table shows these probabilities for k less than 7.

Probabilities of Winning in Ten Roulette Games

Exact number of wins	0	1	2	3	4	5	6
Probability	.002	.015	.059	.143	.225	.243	.182

In particular, the probability of losing all 10 games is about .002. It would be very surprising if she lost all 10 games. Even if she won only 1 game, which has probability about .015, this would be very surprising. Since .059 is pretty small, I'd be rather surprised if she only won 2 games. I wouldn't be very surprised at 3 wins, because this has probability .143. There's about a 1 in 7 chance that she'd win only 3 games.

If we really want to check whether the house is cheating, we should observe a lot more games, so let's suppose we observe 1,000 games. Should we be surprised if she won exactly 474 games? The probability of *exactly* 474 wins in 1,000 games is

$$_{1000}C_{474} \times (.474)^{474} \times (.526)^{526},$$

which turns out to be approximately 0.025. Winning exactly 474 games isn't very likely, but this isn't evidence of cheating. In fact, it turns out that this number of wins is at least as likely as any other. We conclude that *each* number of wins is unlikely, yet there must be *some* number of wins.[2] Nevertheless, although 474 wins wouldn't be surprising, I'd be very suspicious if she only won 200 or 300 games. Why is the observation of 474, or even 470 or 480, wins not surprising, but 200 or 300 is? We can't answer this yet, because we haven't asked for the right probabilities.

The useful definition of p-value turns out to be, roughly speaking, the probability of the observed event *or any more extreme and surprising event*. If our dowager won only 400 games out of 1000, this seems to suggest that something's wrong, so the house is cheating or using faulty equipment. Or does it? This is where we need the p-value. If the roulette game is really honest, we interpret "400 wins or any more extreme and surprising event" to mean the event "400 or fewer wins." So the p-value of the observation "400 wins," assuming an honest game, is

$$\textbf{Pr}(\text{at most 400 wins in 1000 games}).$$

This turns out (using the DeMoivre-Laplace Limit Theorem) to be less than 0.0001. There's less than 1 chance in 10,000 that our dowager would win 400 or fewer games in an honest roulette game.

Here are the p-values for some other observed numbers of wins.

p-values for 1000 Roulette Games

Number of wins	420	430	440	450	460	470
p-value	.0003	.003	.016	.064	.188	.400

---

[2]This is similar to the following remark. Because I don't know your birthdate, each date has very small probability of being your birthdate (≈ 0.003), yet one of those dates must be your birthdate.

Certainly the *p*-value of the observation 470 is not small. Nor would 470 wins be surprising. Its *p*-value tells us that there is about a 40 percent chance that the dowager would win 470 or fewer games. The *p*-value for 460 isn't very small either. On the other hand, the *p*-value for the observation 440 looks small to me. There's only about 1 chance in 62 that the dowager would win 440 or fewer games. The *p*-values for 420 and 430 are even worse.

What about the *p*-value, 0.064, for the observation 450? Is 450 a rare observation? The chances of that many wins or fewer is about 1 in 15.5. What do you think?

In the last two paragraphs, I suggested that some *p*-values were small, so that the observation was surprising, and some were large, so that the observation wasn't surprising. What's surprising, then, depends on what we think constitute "small" *p*-values. It turns out that a convenient cutoff, which is often used by statisticians and users of statistics, is 0.05. That is, we're surprised (and maybe suspicious) about an observation that has *p*-value less than 0.05, and we're not surprised otherwise. As you can imagine, "surprise" is not part of the statisticians' bag of jargon. They often say that the observation is *statistically significant* if the *p*-value is less than 0.05, and not otherwise. What they mean by statistically significant is: Something is probably going on; the observation probably wouldn't have occurred at random. The assumptions (an honest roulette game in my example) may well be false.

In the olden days, *p*-values were not given. Instead, an observation was announced as statistically significant, or not, based only on whether the *p*-value was above or below some cutoff point, which was often 0.05. The advantage of *p*-values is that they allow readers to make their own judgments. The exact cutoff of 0.05 shouldn't be taken too seriously. Obviously, a *p*-value of 0.049 isn't that much more significant than a p-value of 0.051. My rough cutoffs are: If I see a p-value above about 0.08, then I doubt that the observation is statistically significant. If it's between 0.02 and 0.08, then I suspect that the observation is statistically significant, but if it is important (as in a medical study), I would look for more evidence. If the p-value is less than 0.02, I'm convinced that the observation is statistically significant. There's less than 1 chance in 50 that the observation would have occurred at random.

Finally, let's suppose that our dowager actually won 451 games. Is this statistically significant; that is, do we doubt that the game was honest? The p-value of this observation turns out to be about 0.073. In the olden days, statisticians would say that this is not statistically significant because it is bigger than 0.05. The dowager was just a bit unlucky. But we can use our own judgment. Using my guidelines, I admit that the result is not statistically significant, but it's close enough to the cutoff value to make me nervous. I would move to another roulette table.

This long roulette example illustrates how p-values are used when one needs to assess whether some observation suggests that some assumptions may not be true. If the p-value is sufficiently small, then the observation is statistically significant, and we can be quite sure that something is wrong with our assumptions. If the p-value is not small, then the observation is not statistically significant. We cannot conclude that our assumptions are correct, but we can say that the observation is compatible with the assumptions. One might even say it supports the assumptions, but that's dangerous language because it does not confirm them.

Finally, I return to baseball! Baseball fans, with a statistical bent, have asked questions like these: How surprising is it that someone had a 56-game hitting streak in the past

100 years? Should we be surprised at how often particular batters hit in the clutch? We'll return to questions like these. Frequently, the answers will be expressed in terms of p-values. Here's some very specific questions that I can answer now.

In 2003, Texas shortstop Alex Rodriguez's batting average was .298. His lifetime *AVG* before 2003 was .309. Is this surprising? Was he in a slump? Or could this have just been due to random bad luck? I admit that this is a strange question considering that he won, and deserved, the Most Valuable Player award in 2003.

Alex's 2003 average is based on his 181 hits in 607 official at-bats. Methods similar to those for analyzing roulette games show that the probability that he would have had 181 or fewer hits in 607 at-bats, if his at-bats were Bernoulli trials, is ≈ 0.28. This is not statistically significant. In other words, even if Alex had been a robot with probability of success .309 at every at-bat, the chances are quite good that he'd have hit .298 or less in 2003. This was no slump, just a little bit of random bad luck.

Cal Ripken's *AVG* was .340 in 1999. His lifetime *AVG* before 1999 was .276. Is his .340 average in 1999 surprising? Or might his *AVG* have jumped this much just by chance?

What we need is the p-value of this observation if, in fact, Cal's batting performances were Bernoulli trials with $p = 0.276$. This will give us an approximation for how likely it is that he would bat .340 just by the laws of chance. Cal's 1999 season consisted of 332 trials (official at-bats) and he had 113 successes (hits). In this setting, it would be more extreme and surprising if Cal had hit *more* than 113 hits. So the p-value is the probability that Cal would obtain 113 or *more* hits in 332 official at-bats, which turns out to be ≈ 0.0052. This p-value is very small, so it is very unlikely that Cal's improvement was just random luck. Something changed. In fact, 1999 was the first year that Cal didn't play every day. It seems evident that his hitting improved substantially when he had days off.

In 2003, Yankee center-fielder Bernie Williams' batting average was .263, though his lifetime average before 2003 was .308. Is this statistically significant?

In 2003, Bernie had 117 hits in 445 official at-bats, which gives $AVG = .263$. If he were a robot with probability of success .308 at every at-bat, the probability that he would have had 117 or fewer hits in 445 official at-bats is ≈ 0.02. Most people would agree that this is statistically significant. Bernie was in some sort of slump. Something had changed in Bernie's hitting for the year 2003. Whatever it was, he recovered for the World Series, when he had a .400 batting average.

In the roulette example, the repeated betting on red numbers really would be a sequence of Bernoulli trials, assuming that the casino is honest. However, in the batting-average examples, I assumed that the repeated events were approximately Bernoulli trials, even though I know they are not Bernoulli trials. For one thing, the probability of a player's success (a hit) isn't the same at each at-bat; it varies depending on the pitcher and many other factors. Assuming that the at-bats are Bernoulli trials is a useful simple approximation to get a feel for the relevant p-values.

For many statistical situations, it is not appropriate to even consider Bernoulli trials as a model. I give an example.

The 1998 Yankees was one of the greatest teams of all time. Their record of 114 wins and 48 losses was spectacular; their winning percentage was .704. Their performance in the postseason was even more impressive. They won 11 of 13 games, and four of the wins were shut-outs. During the season, they outscored their opponents 965-656, which averages out to 6-4 each day. In the postseason, they outscored their opponents 62-34, so they scored 64.6 percent of the runs in those games. I wondered whether they were better, i.e., significantly more effective, during particular periods of the games. The three periods I focused on were the first three innings, the next three innings, and the remaining innings. If the Yankees were equally effective during each period, we would "expect" them to score about 64.6 percent of the runs scored in each period of the games. In the next table, I give the actual number of runs in each period, followed by the expected number of runs per period. Because the Yankees and their opponents scored 35 runs in the first three innings, the expected number of Yankee runs during those innings should be 64.6 percent of 35, which is about 22.6. I subtracted this from 35 to get the opponents' expected number of runs in the first three innings. The other expected numbers were obtained in the same way.

The question is whether these data suggest that the Yankees were more effective during some periods than other periods of the games. From the table, it looks as if they were unusually effective the first three innings and not very effective during the middle innings. But is this obser-

Runs Scored in 1998 Yankees' Postseason Games

	Innings 1–3	Innings 4–6	Innings 7 on	Total
Yankees' actual	29	16	17	62
Opponents' actual	6	19	9	34
Total actual	**35**	**35**	**26**	**96**
Yankees' expected	22.6	22.6	16.8	62
Opponents' expected	12.4	12.4	9.2	34

vation statistically significant? It's not reasonable to try to model these data as Bernoulli trials. What's success? What are the independent trials? What's p? Nevertheless, there's a way to find a p-value for the observations in the table. It turns out that there's a way to measure how far each possible array of values in the first two rows of the table is from the array of expected values (in the last two rows of the table). Each possible such array, with row totals 62 and 34, is assigned a positive number. The bigger the positive number, the more unexpected the array is. The p-value of an observed array is the probability of this array or any other arrays that are more unexpected.

It turns out that the p-value of the Yankees' actual runs scored, as given in the table, is about 0.06. Using the usual cutoff of 0.05, I could conclude that the observed deviation is not statistically significant, but the p-value is so close to 0.05 that I choose to view it as statistically significant. This raises the question of what specific parts of the observed data are statistically significant. I already gave my suspicions based on a glance at the table (unusually effective in the first three innings, etc.). However, a careful analysis to confirm my suspicions would require more advanced statistical methods.

I also wondered whether the situation was different for the 2003 World Series' champions, the Florida Marlins. This team was very different from the powerful 1998 Yankees, being a wildcard team that was an underdog throughout the postseason. In fact, the Marlins were (barely) outscored by their opponents, 77–79. They scored 49.36 percent of the runs scored in those games. The next table gives the actual and expected values for their 2003 postseason. Because the data in this table seem less extreme than the data in the Yankees' table, I don't expect these data to be statistically significant. And I'm correct; the p-value

Runs Scored in 2003 Marlins' Post-season Games

	Innings 1–3	Innings 4–6	Innings 7 on	Total
Marlins' actual	23	26	28	77
Opponents' actual	35	23	21	79
Total actual	**58**	**49**	**49**	**156**
Marlins' expected	28.6	24.2	24.2	77
Opponents' expected	29.4	24.8	24.8	79

turns out to be about 0.61. The observed differences from the expected values are definitely not significant.

The method used for finding p-values from the two tables just mentioned is called a *chi-square test*. The positive numbers assigned to arrays are written as χ^2 (that's a Greek chi), and the p-value is the probability that χ^2 is as large as observed or larger. Fortunately, statistical packages and spreadsheets do the calculations for us.

Do players in various sports get in "the zone" or "groove," so that they have streaks of high performance beyond what we'd expect? This is a difficult question and it is rather ambiguous. So, let's start with a more specific question: Do batters have hitting streaks?

The answer to this question is: Of course! We all know of examples, and some record streaks are well known. The most famous hitting streak is Joe DiMaggio's streak in 1941, when he got at least one hit in each of 56 consecutive games. Because few players have even had 40-game hitting streaks, most fans regard DiMaggio's record as remarkable and truly unexpected. Many doubt that this record will ever be broken. Some exuberant DiMaggio fans might even call this record a miracle, but in fact a lot of hard work and skill was involved. In any case, several statistically-minded fans have wondered just how unlikely such an event is. Let's be clear which event we're wondering about. Consider the following questions.

1. How likely was it that someone would have a 56-game hitting streak in the past 100 years?
2. How likely was it that Joe DiMaggio would have a 56-game hitting streak sometime in his career?
3. How likely was it that Joe DiMaggio would have a 56-game hitting streak in 1941?

Clearly, each possibility was less likely than the preceding, and we would expect the probabilistic answers to Questions 2 and 3 to be low. What about Question 1? If that probability is low too, then DiMaggio's record really is remarkable. If that probability is high, then the record isn't that amazing and someone has to have the record. And DiMaggio is certainly a reasonable candidate to have this record.

So, batters have hitting streaks. But does this mean that batters are in a groove? Or in "the zone?" We wouldn't think of robots or computers as being in the groove or zone. Is there something different about the batters' performances that is creating the streaks? More interesting and intriguing to statisticians: Do some players have more and longer streaks than one would expect if they were consistent like a robot or computer program?

I haven't done research on this interesting question, but some statisticians have. What follows is what I have learned from reading several of their papers, which are listed at the end of Appendix 1, starting on page 165.

Christian Albright wrote a frequently-cited paper for the *Journal of the American Statistical Association* in 1993 that provoked a response from Jim Albert in the same journal. The basic question was whether hits or wins, say, appear in a pattern that is different from the pattern that would be obtained in a purely random situation.

Albright analyzed hits for the years 1987–1990 for regular players, i.e., players who had at least 500 plate appearances. There were about 125 such players each year, and 40 of them played regularly all four years. Albright looked for "streakiness" and its opposite, "stability," i.e., less variation than one would expect. He used three methods to look for comparisons with random data.

1. He counted the number of streaks, including streaks of length 1, over a season, and then he checked to see how these numbers deviated from what would be expected if the sequence of hits and nonhits were completely independent, i.e., if they were Bernoulli trials. Statisticians call this a "runs test" because they call streaks "runs." We don't, because the word "run" is already taken. As an example, the following string of hits H and nonhits N

   H H N N N H N N N N H N N N N H H N H

   has nine streaks, including two streaks of four nonhits. *p*-values of the number of observed streaks can be obtained to determine how likely it is that the streaks occurred pretty much at random. Note that a streaky player would have unusually long streaks, so the total number of streaks would be *lower* than expected. An unusually *high* number of streaks would suggest that the hitter was even more regular than expected, which Albright called "stability."

2. The second method is based on counting the number of changes of each type, $H \rightarrow H, H \rightarrow N,$ $N \rightarrow H,$ and $N \rightarrow N.$ In the string, analyzed in method 1, there are eight changes of type $N \rightarrow N.$ These data can be tested using a chi-square test.

3. The third method is a very comprehensive "linear regression model with explanatory variables" in which the author takes into account eight varying conditions: (1) whether the game is home or away; (2) whether the game is a day game or a night game; (3) whether the game is played on grass or artificial turf; (4) whether the pitcher is left-handed or right-handed and the pitcher's current-year earned-run-average; (5) the number of runs his team is ahead or behind; (6) whether there are two outs, (7) whether there are runners on base or in scoring position; and (8) whether the game is in the first six innings.

Albright worked with plate appearances, rather than at-bats, but he indicated that the results are about the same either way. As I've already emphasized, with lots of data, there's sure to be some streaky-looking data, even if the process is completely random. In fact, the problem is deciding whether there's more or less streaky data than randomness would dictate. Here are two interesting extremes from this paper.

In 1988, Dwight Evans was very streaky. His *OBP* was .384. The expected number of streaks in 628 at-bats was 298, but he only had 261 streaks. And his batting average was .124 higher after a hit than after a nonhit. In 1989, Harold Reynolds was very stable, i.e., very nonstreaky. His *OBP* was .367. The expected number of streaks in 667 plate appearances was 311, but he had 346. His batting average was .115 lower after a hit than after a nonhit. But these patterns didn't even occur for these players for other years. If Evans was intrinsically a streak hitter, and Reynolds was intrinsically a stable hitter, we'd expect these intrinsic skills to show up each year.

These cases are interesting, but they are not the real issue. The question is: Is there overall streakiness? Albright concluded that: "The data analysis performed here . . . has failed to find convincing evidence in support of wide-scale streakiness. In fact, the evidence is more in line with a model of randomness."

Jim Albert is dubious. He feels that Albright dismisses the individual's streaky patterns too quickly. He argued that there was evidence that some players' hitting was streaky. From his own study with 200 players in 1989-1990, he found that the most significant "varying condition" was the pitcher's earned-run-average, i.e., how good the pitcher was. He also observed that batters hit better (a) with runners on base, (b) playing at home, (c) against pitchers of the opposite hand, and (d) with less than two outs.

More recently, Jim Albert and Patricia Williamson co-authored a paper in which they used simulation to analyze streakiness, with a focus on two interesting cases. First, they wondered whether Mark McGwire's home run hitting was streaky. After a detailed study using observed five-game home run counts during 1995-1999, they concluded that there is "little evidence that McGwire's recent pattern of home run hitting is different from the coin-tossing model." Indeed, using their data the p-value is about 0.677. Then they asked whether Javy Lopez was a streak hitter in 1998, because there were commentators who said he was. Just looking at the data, it appears that he was a consistent hitter for the first 40 games, hot the next 35 games, and then oscillated between hot and cold. Still, after studying six statistics, Albert and Williamson concluded that they do not provide support for the belief that Javy was a streak hitter in 1998. There is still no solid evidence for streakiness.

Jim Albert grew up in Philadelphia idolizing Mike Schmidt, who was famous for being a consistent ballplayer. In another paper, Albert tested Schmidt's consistency in hitting home runs. Among other data, he observed the information given in the following table.

Mike Schmidt's Games with Multiple Home Runs

Home Runs in a Game	Expected Times	Schmidt's Times	p-value
2	42	43	0.50
3	2	2	0.60
4	0	1	0.04

The last two rows are really meaningless; the numbers are too small. In any case, Albert created and simulated a "consistent" home run hitter model for Schmidt. He focused on the years 1974-1987. In those years, Schmidt's best year had 48 home runs and his worst year had 21 home runs. The 48 was not surprising, but the 21 was low with a p-value of 0.03. Using two-week groupings, Albert concluded that Schmidt had more hot and cold periods than predicted with the consistent model. Schmidt became more consistent in his later years.

Paul M. Sommers used the basic "runs test" to analyze the number of streaks in the home run records of some great hitters. He confirmed that Mark McGwire's home run hitting appears random. He went on to show this was also true of Sammy Sosa, Roger Maris, and Babe Ruth.

Scott M. Berry used methodology similar to that of Jim Albert to analyze the home run records of Mark McGwire, Sammy Sosa, and ten other good home run hitters in 1998, including Barry Bonds. He concluded that only Sosa appeared to have a streaky year. He also noted that Andres Galarraga had a nonstreaky year, i.e., what Albright would call a stable year.

Scott Berry wrote: "I am not a believer in the hot hand. Sure, I believe that a player's home run ability changes throughout a season. Many factors affect a player's home run ability. The handedness of the pitcher, the weather, the ballpark, and the batter's health clearly have an effect on the batter's ability to hit a home run. . . . It would be interesting, and I think the strongest proof available of the existence of streaks, if an analysis could be done in which a model that incorporates streaks could be shown to predict future observations better than a model that did not have a streak factor. I conjecture that the best opportunity for this to be done is in golf."

I agreed with Berry for many years, not in my heart, but because the research papers did not provide strong evidence that there are real streaks in baseball. But the last sentence in Berry's conclusion is interesting. Golf would be a better setting for such a study than baseball or basketball. But bowling or horseshoes would be even better, because the competitors are subjected to more constant conditions than in the other sports.

Here, finally, is an article that contains convincing evidence that there are streaks in . . . professional bowling. The article is, "Bowlers' Hot Hands," by Reid Dorsey-Palmateer and Gary Smith. They justify their choice of bowling as follows: "Bowling data are relatively clean in that, unlike basketball data, every roll in a game is from the same distance and made at regular, brief intervals."

Basketball free-throw data, which I mention in Chapter 6, are pretty clean, but generally lack the "regular, brief intervals" feature. Dorsey-Palmateer and Smith analyze frame-by-frame results from the 2002-2003 Professional Bowlers Association (PBA) tour for all match-play games, which are available from the PBA.

The Dorsey-Palmateer/Smith article found that, for many bowlers, the probability of rolling a strike is *not* independent of previous outcomes. Moreover, the number of strikes rolled varies more across games than can be explained by chance alone. In particular, most bowlers have a higher strike proportion after 1, 2, 3, or 4 consecutive strikes than after the same number of consecutive nonstrikes. This difference becomes more pronounced as the number of consecutive strikes increases. Here's some of the data:

Strike Proportions Immediately Following Consecutive Strikes and Nonstrikes

Number of Consecutive Strikes/Nonstrikes	1	2	3	4
Number of bowlers	134	111	81	43
Proportion of strikes, after strikes	.571	.582	.593	.612
Proportion of strikes, after nonstrikes	.560	.546	.533	.492
p-value of differences of proportions	.020	.000	.000	.000
Bowlers with higher proportion after strikes	80	77	59	34

The p-values in the table are rounded to 3 decimal places; none of them is exactly 0, of course. These very low p-values are convincing evidence that bowlers get in "the groove." The last row shows that most bowlers roll more strikes after consecutive strikes than after consecutive nonstrikes. For example, 34 of the 43 bowlers had a higher percentage of strikes after 4 consecutive strikes than after 4 consecutive nonstrikes.

Dorsey-Palmateer and Smith also note that there were 19 perfect games (12 strikes for a total of 300 points) on the 2002-2003 PBA tour. Based on 100,000 computer simulations of the tour, using each bowler's strike proportion and assuming their rolls are Bernoulli trials, they found that there's about a 3 percent chance that there would be 19 or more perfect games, i.e., the p-value of 19 perfect games is 0.03. This is further evidence that bowlers get in the groove.

Early papers (1985-1989) by T. Gilovich, A. Tversky, and R. Vallone use statistics to debunk the common perception that basketball players sometimes have "hot hands." But Dorsey-Palmateer and Smith argue that their data are flawed and, as Jim Albert has also argued, statistical tests often have little power to detect hot hands. Dorsey-Palmateer and Smith go on to argue: "Unfortunately, their data do not identify how much time passed between shots. A player's two successive shots might be taken 30 seconds apart, 5 minutes apart, in different halves of a game, or even in different games. Another problem is that the shots a player chooses to take may be affected by his recent experience. A player who makes several shots may be more willing to take difficult shots than is a player who has been missing shots. In addition, the opposing team may guard a player differently when he is perceived to be hot or cold. Shot selection may also be affected by the score and the number of fouls accumulated by players on both teams."

After years of acknowledging that streakiness seems to be merely random, I must now assert what I have long intuited: *Sports players do get in "the groove" or have "hot hands," but this intrinsic human quality is very difficult to identify in complex games like baseball and basketball.*

As time goes on, statisticians will undoubtedly develop better tools and use more relevant data for questions about streaks. The more they look, the more we'll know. But I personally doubt that we'll ever have definitive answers to questions about streaks in baseball, simply because it's impossible to take into account all the human elements. People are much more difficult to study and understand than robots.

Back to Joe DiMaggio's hitting streak for 56 games. Michael Freiman took an interesting approach to the subject. For many ballplayers of the past, and specified years of their careers, he calculated the probability that they would have a 56-game hitting streak in the specified year. Of the 45 players with the best chance, a majority played during the 19th century, and no player made the list after 1930. The top 45 players had probabilities between 0.0328 and 0.0018, whereas Joe DiMaggio's in 1941 was only 0.0001. Of course, after the fact, all these probabilities are meaningless. Freiman indicated that the probability of such a streak before 1931 was about 0.36, while the probability of such a streak since then was only about 0.05. "Thus the most surprising part of the DiMaggio streak may not be that it happened at all, but that it happened so late in the history of baseball."

In private communication, my friend Dan Schlewitz took a different approach to Joe DiMaggio's streak. To determine how likely DiMaggio's streak was, he created 100 computer simulations of the 1941 season, using the baseball computer game, DiamondMind Baseball. The five longest streaks in that 100 years of 1941 baseball were 55, 52, 40, 39, and 36, the last three of which belonged to DiMaggio. All told, DiMaggio had the longest hitting streak in 16 of the 100 seasons. In any case, these results suggest that DiMaggio's 56-game hitting streak was remarkable, but that very long streaks are to be expected every few decades. It is interesting to note that the Yankees won the pennant in 88 of these "1941 seasons." In reality, they won the pennant in 1941, *seventeen* games ahead of the Boston Red Sox, even though Boston's team *AVG, OBP,* and *SLG* all exceeded the corresponding team numbers for the Yankees.

Keith Karcher wondered whether teams have winning and losing streaks more than would be expected if their wins and losses were independent, i.e., Bernoulli trials. Using "runs tests," i.e., counting winning and losing streaks, he concluded that there were generally no more streaks than if the outcomes were Bernoulli trials.

Because an oddsmaker suggested that one could make money by betting on a win after three consecutive wins, and betting on a loss after three consecutive losses, Karcher also analyzed such streaks. That is, he analyzed the likelihood of a win after three consecutive wins and the likelihood of a loss after three consecutive losses. He found no evidence that the oddsmaker's scheme would make money.

The mystery of streaks is especially difficult for mathematics to explain and fascinates statisticians and baseball fans alike. I've concluded that sports figures tend to have streaks in whatever they are doing, but this is difficult to verify in general. Appendix 1 is a report on some other research papers that tackle easier, but nevertheless interesting, issues.

My second love, probability, is the beautiful theory that makes statistics work. And statistics is the tool that helps us make sense of data in the world. It's the tool that helps us decide when certain observations are so rare that we suspect that something beyond randomness is going on. But, from the point of view of statistics, the real world and the people in it are messy, i.e., hard to pin down. My first love, baseball, is a great example of this.

Appendix

1

Now a Word from Our Statisticians

Both amateur and professional statisticians and mathe-maticians have published many interesting articles about baseball. I will mention a few journals and comment on a few articles that I found especially interesting. At the end of this appendix, I will discuss a recent collection of essays involving statistical analysis of baseball.

The Society for American Baseball Research (SABR) has over 6600 members worldwide. The primary areas of research are the history of baseball and statistics in baseball. Of course, there's a natural overlap of these two areas.

In addition to publishing many books, SABR publishes several journals. One is *The National Pastime: A Review of Baseball History*. This journal is published about once a year. Each issue contains numerous short articles, often on esoteric topics. Statistics is not the major emphasis, but they inevitably creep in. Issue number 23, published in 2003, contains 23 articles (surely a coincidence), including "The Spitball and the End of the Deadball Era," "The Statistical Impact of WWII on Position Players," "Ted Williams in 1941," and an article by Alan Abramowitz titled "Is There a Home Field Advantage in the World Series?".

Another SABR journal is *The Baseball Research Journal.* Issue 31 was published in 2003. Some of the intriguing titles of the typically short articles are: "Batting Average by Count and Pitch Type," "Baseball's Most Unbreakable Records," and "The Best Last-Place Team Ever." The first article will be discussed in this appendix. The second article gives a long list of records, some quite obscure, that the author argues are least likely to be broken. Among them are Carl Yastrzemski's lowest league-leading *AVG* of .301 in 1968, Ty Cobb's highest career *AVG* of .366, and Rogers Hornsby's highest five-season batting average of .402. Yankees-haters may be annoyed to learn in the third article that the Yankees "win" here too, in 1966. I eagerly await an article on "The Worst First-Place Team Ever."

An economist, Paul M. Sommers, has written several articles, many jointly with other authors, about the correlation between salaries and performance and similar correlations related to economic issues. He also wrote the article, "Home-Field Advantage in the World Series: Myth or Reality?" Using *p*-values and chi-square tests, he shows that since the adoption in 1924 of the familiar 2-3-2 best-of-seven format for World Series, there's a significant home-field advantage for games 1, 3, and 6 and only for those games. In other words, there's a home-field advantage for the first game of each home stand, but not for the other games. Abramowitz's article in *The National Pastime,* which I mentioned earlier, only addresses the question: Is there an advantage for a team to play the first two

World Series games at home? He concludes that there is a significant advantage, because the teams with this advantage have won 58 percent of the time.

The article "Eye-pothesis Testing: Another Look at the 1986 World Series," by Andrew Cornish, Charles Fiedler, Brian S. Foss, and Paul M. Sommers, concerns whether batters have a "better eye" when faced with two strikes. The authors analyzed the seven-game 1986 World Series, between the Boston Red Sox and the New York Mets, pitch-by-pitch. The assumption is that a batter with a "perfect eye" would never take a called strike, though in practice some batters will deliberately take first pitches, whether they are strikes or not, and they will also take called strikes when they have less than two strikes and they feel the pitch is not good enough for them. In any case, with less than two strikes, pitches taken turned out to be called strikes 36.9 percent of the time, 325 out of 880. With two strikes, pitches taken only turned out to be strikes 13.3 percent of the time, 26 out of 196. The observed difference here is extremely statistically significant. In fact, the p-value for such a difference to occur at random is about .00000000008. Batters are *much* more careful about taking pitches when faced with two strikes. There's absolutely no surprise here. But it is still interesting to see this well-known fact confirmed statistically, even though the analyzed data are rather specialized, being based on a single World Series. The authors also found that, for this World Series, the batters of the two teams did not have statistically significant different "good eyes" while playing in New York. However, Boston batters had statistically significant "better eyes" than New York batters while playing in Boston, since the p-value $\approx$.003.

Similar issues are addressed by J. Eric Bickel and Dean Stotz in "Batting Average by Count and Pitch Type: Fact

and Fallacy." They cite two authors and a sports announcer who make strong arguments that batters should
always swing at good pitches before they have two strikes,
because statistics show that batting averages are lower
when calculated for batters with two strikes. In fact, batters' averages are between .070 and .170 *lower* when they
have two strikes. Yet many successful batters deliberately
take first pitches, as advised and practiced by one of the
best, Ted Williams. And most batters are pickier about
what they'll swing at when they have less than two strikes.
What's going on here? The authors point out the fallacy in
looking at these batting averages based on pitch count.
"There are *four* ways to have an official at-bat with less
than two strikes: hit, error, fielder's choice, or batted out
(the ball has to be put in play). However, there are *five*
ways to have an official at-bat with two strikes: hit, error,
fielder's choice, batted out, *and* strike-out." Thus, the AVG
when players complete their at-bats with *less than* two
strikes is the fraction

$$\frac{H}{H + E + FC + BO},$$

where the numbers H, E (errors), FC (fielder's choice), and
BO (batted out) are the numbers for less than two strikes.
The AVG when players complete their at-bats with two
strikes is

$$\frac{H}{H + E + FC + BO + SO},$$

where all the numbers are based on two strikes. What is
misleading is that all the strike-outs, abbreviated SO, appear in the second fraction. This significantly increases the

denominator, and hence lowers the observed average. Using a large database for college baseball played by Stanford and competing teams, the authors calculated *AVG* and *SLG* based on each possible pitch count. They found that the chance of getting a hit with two strikes was at least as high as every other pitch count, except for 0 and 2 (0 balls and 2 strikes). The statistics cited by the two authors and sports announcer are not wrong, but they are misleading! For a better analysis, the authors propose two new statistics,[4] *HPS* (hits per strike) and *IPA* (probability of a hit given that the ball is put into play).

Bill James has created a new system for evaluating all ballplayers, called "win shares." This complicated concept is explained in the erudite and interesting book [26, *Win Shares*] by Bill James and Jim Henzler. Win shares take into account various aspects of fielding as well as hitting, and pitchers are also allocated win shares. Moreover, win shares are adjusted to take into account era, ballpark, and other factors. In short, win shares are designed to evaluate the overall value of players based on their contributions to *wins*. The idea is this. For a season, the win shares for each team is 3 times the number of games that the team won that season. The choice of 3 is explained on page 2 of [26]. Based on an extremely elaborate set of rules, these team win shares are apportioned to all the players on the team.

The book [26, *Win Shares*] is a tough read, and it has no index. The book [25, *New Bill James Historical Baseball Abstract*] is a much friendlier book, with an index. The section "Player Ratings and Comments" on pages 329–369 includes an explanation of win shares. James ends this section with his list of "the 100 greatest players of all time." As he readily admits, and discusses in some

---

[4]Trademarks of Competitive Edge Decision Systems.

detail, fans will find his ratings controversial. Pages 927–970 include tables of players' win shares listed by year and by the decades in which they were born. Win shares are also given for players on a cross-section of 24 teams, from the 1953 Brooklyn Dodgers to the 1999 Tampa Bay Devil Rays.

In 2003, SABR published, *The Best of By The Numbers, a Collection of Thought-Provoking Essays on Baseball by SABR's Statistical Analysis Committee.* Edited by Phil Birnbaum, this collection consists of 17 short essays that originally appeared in the SABR journal, *By the Numbers,* in the period 1989–2001. Each article is preceded by a brief introduction, written by members of SABR's Statistical Analysis Committee. Below, I will give a thumbnail sketch of each article. I hope some of you will be so intrigued that you'll join SABR.

I should emphasize that broad summaries aren't substitutes for the original articles. For one thing, I tend to simplify the conclusions and ignore subtleties that the authors discuss. For another thing, results in many of the articles suggest behavioral changes for players and managers. But in the real world, actions lead to reactions. Changes in offense lead to changes in defense. Changes in hitting style lead to pitching adjustments. And so on. So, the authors usually caution that their conclusions cannot take into account other, sometimes unexpected, side effects of a possible strategic change.

Here are the promised summaries.

"Clutch Hitting One More Time," by Pete Palmer. Does clutch hitting exist? Of course. All fans know players who've been very successful in "clutch" situations. We know some who've been very unsuccessful too. The real question is this: Are some players *intrinsically* better, or worse, hitters in "clutch" situations than in other situations? Or are they just lucky, or unlucky? Because of the publicity given to players who excel in these situations, and some who seem to do poorly under pressure, most fans believe the answer to the question is Yes. However, this conclusion does not hold up under close statistical scrutiny! The evidence is that clutch performance is not a skill. Palmer compares the clutch batting averages in late innings with the nonclutch batting averages for the 330 most active hitters of the 1980s. When the variations are normalized,[5] one would expect a picture of the results to be close to the bell curve *if the process was random*. If the process was not random (i.e., if some of the hitters had the "skill" to do better, or worse, in clutch situations), we would not expect the bell curve effect. In fact, the results fit the bell curve quite well. There were no surprises. This is why we statistical types believe that clutch hitting, beyond good or bad luck, does not exist! Also, see Chapter 9 on "Measuring Clutch Play" in [2, *Curve Ball*], which discusses a variety of measurements for clutch plays. A list of the ten top "clutch moments in baseball" is given.

"Does a Pitcher's Stuff Vary from Game to Game?" by Phil Birnbaum. If so, one would expect to be able to use a pitcher's performance early in the game to predict how well he will do later in the game. It turns out that a bad first inning (giving up 3 or 4 runs, or giving up 2 or 3 walks) generally had a small effect on the pitcher's performance for the remainder of the game. This was surprising to the author! However, the author offers some cautions in interpreting the data. One is that his data necessarily include only the results *when the pitchers are*

---

[5]Normalizing a statistic involves multiplying and/or adding constants to be able to compare the statistic with some standard statistic.

allowed to pitch beyond the first inning. Perhaps, when some pitchers lose their stuff, the manager or pitching coach can "see" it and so the pitcher is removed from the game. The author seems doubtful, though, that these possibilities are significant enough to change his conclusions.

"Catchers: Better as Veterans," by Tom Hanrahan. The author addresses the question of whether catchers get better at some of the subtler aspects of their job behind the plate, such as handling pitchers and framing the strike zone, as they age and gain more experience. As the title of the article suggests, catchers' defensive statistics generally improve as they gain experience. No surprise here, but it's always nice to get our intuition confirmed.

"How Often Does the Best Team Win the Pennant?" by Rob Wood. The short answer is, "a little less than half the time." This study focuses on pennant winners over the past 100 years, not on postseason outcomes. To simplify comparisons of the olden days and recent years, the author ignores division outcomes and defines the pennant winner, in each league, as the team with the most wins during the regular season, so ties are possible. The best team is determined using a concept called the team's "innate winning percentage." Up until about 1962, the best team won the pennant about two-thirds of the time. Since then, the percentage of best teams winning the pennant has dropped to under 50 percent. The author compares this with his view that the best team regularly wins the Super Bowl, the NBA championship, and the Stanley Cup. He offers some explanations for these interesting observations. Chapter 11, "Did the Best Team Win?" in the book [2, *Curve Ball*] uses different methods and relies on computer simulations, but reaches similar conclusions. In fact, the authors of [2] figure that the best baseball team will win the pennant with about 0.35 probability.

"Playing Every Day and September Performance," by Harold Brooks. Recall my comparison in Chapter 8 of Cal Ripken's *AVG* before and after he stopped playing every day. This article was written in 1992 when Ripken was still playing every game, so the author begins by comparing Ripken's performance before September 1, over the years 1984–1991, with his performance in September and October over those same years. For example, his *AVG* before September 1 was .283, while for September and October it was .255. Ripken's other statistics, such as *SLG,* showed similar declines after September 1. The author goes on to show that many players, who play nearly every day, exhibit similar tendencies, though usually not so extreme. A striking observation is the poor late-season performance of throwing infielders (2B, SS, 3B) who get fewer than five games off in a season. With more rest, these infielders have a much greater chance of doing well in the latter part of the season.

"Does Good Hitting Beat Good Pitching?" by Tom Hanrahan. This is a popular question, as is its counterpart, "Does Good Pitching Beat Good Hitting?" These questions are tougher to nail down than most questions that SABR folks tackle. Hanrahan rephrased the questions this way: *When superior batters face superior pitchers, are the results different from expectations predicted by using typical mathematical models of their abilities?* The author approaches the question in two distinct ways, but his final conclusion is that good pitching does not beat good hitting, or vice versa, any more than we should expect. Bill James is convinced, and so am I, even though many people believe that good pitching beats good hitting.

"Stolen Base Strategies Revisited," by Tom Ruane. This follows up on earlier work regarding the value of the stolen base. Is the prospect of improving the likelihood of manufacturing runs worth the risk of making an out? It has been generally agreed that at least a two-thirds rate of success is needed to improve scoring prospects. But this general rule ignores special situations, for example, where only one run is needed, or when there are two outs, or when there's a weak hitter at the plate. This paper provides a more detailed analysis of the most common situation: There's only a man on first. Should he attempt to steal second base? Various circumstances, such as the number of outs, are analyzed.

"The Recipe for a Stolen Base," by Sig Mejdal. A list of ingredients is easy: the base runner, catcher, pitcher, umpire, and playing surface. This article tells us their relative importance for a successful stolen base. The most important ingredient is the base runner. No surprise, so far. But the pitcher is nearly as important, while the catcher's skill is substantially less important. Playing surface matters some, but the umpires don't, statistically speaking.

"Do Faster Runners Induce More Fielding Errors?" consists of two short articles, one by Dan Levitt and one by Cliff Blau. Intuition and experience certainly convince me that the answer is Yes. Based on 1996 and 1997 data, Levitt found that team speed correlates somewhat with opposition errors and unearned runs. He concludes that there's a "positive but not overwhelming relationship between team speed and opposition errors." Using data supplied by Tom Ruane, covering 1980 to 1998, Blau observes that the average player reached a base on error about 14 times per 1000 at-bats. Blau wondered whether this average was significantly different for different groups of players: left-handed players, home run sluggers, hitters who hit into double-plays a lot, and so forth. There are observed differences from group to group, but Blau concludes that they are not statistically significant.

"Finding Better Batting Orders," by Mark D. Pankin. I'll just summarize this sophisticated article. The difference between using a typical conventional batting order and using an ideal batting order, based on the author's research, turns out to be only one victory per season. In general, then, it isn't worth agonizing over perfecting a batter order. On the other hand, some managers could benefit from realizing the importance of *OBP* for the number one and number two batters. For the lead-off batter, and I quote the author, "getting on base is everything. To a much lesser extent, home run hitters should not lead off. Stolen base ability is irrelevant. . . . I never cease to be amazed by managers who are so fascinated by speed that they forget players can't steal first base!"

"Offensive Replacement Levels," by Clifford Blau. Bill James ordered fielding positions in order of their importance for defense, called the "defensive spectrum." Here it is

$$1B, LF, RF, 3B, CF, 2B, SS.$$

As we all know, the most critical defensive fielding positions are up the middle, *SS, 2B,* and *CF.* Blau analyzes the level of hitting needed for players to keep their jobs, i.e., the "offensive replacement level." We would expect this level to be highest for players at first base and lowest for shortstops. Blau's analysis yields numbers to confirm this general expectation. As one example, for players whose runs created per 27 outs were modest (3.5 to 4.0), the percentages of regulars who returned as regulars the following season *by position* were:

Percentages of Regulars Who Returned as Regulars

Position	1B	LF	RF	3B	CF	2B	SS
Percentage	48.4	40.0	55.6	67.2	66.2	80.2	78.4

The weaker offensive players at second base and shortstop were substantially more likely to keep their jobs than their offensive counterparts at first base or in left field.

"A New Way of Platooning: Ground Ball/Fly Ball," by Tom Hanrahan. Experience has shown that right-handed batters hit better against left-handed pitchers, and vice versa. This fact is used a great deal by managers. Are there other general observations that are equally significant and worthy of managers' attention? Using correlations, the author shows that ground-ball hitters do better against fly-ball pitchers, and fly-ball hitters do better against ground-ball pitchers. The effects are larger for batters who strike out often. In a few cases, the effects are larger than the standard left-right differences. Unfortunately, a new complication is that the definitions of "ground-ball hitters" and "fly-ball hitters" reflect matters of degree, and many hitters aren't classifiable as either. There's a similar problem for pitchers. Hanrahan's new insight won't help unless the hitters and pitchers have strong tendencies to fit the descriptions. The author provides a very convincing example, though. Mark McGwire was the most extreme fly-ball hitter with a lot of strike outs. His *SLG* and *OBP* were .721 and .468 against ground-ball pitchers and only .624 and .392 against fly-ball pitchers.

"The Run Value of a Ball and Strike," by Phil Birnbaum. When umpires call strikes, sometimes they may be influenced by how catchers frame pitches. Birnbaum wonders what fraction of a run is saved, on average, when an umpire calls a strike when the pitch was actually a ball. On average, coaxing a call strike on a pitch that was a ball saves the defense about 1/7 of a run. The effect adds up. If a pitching staff obtained one extra call strike per game in a season, that should result in about 20 more runs and about two extra wins.

"Average Run Equivalent Method," by Clem Comly. The author uses this new method, which he calls ARM, to analyze the ability of outfielders to shut down the opponents' running games. ARM takes into account assists and also errors that allow base runners to take extra bases. The best regular outfielders save about 10 runs per season, and the worst ones allow about 6 extra runs per season.

"What Drives MVP Voting?" by Rob Wood. This article confirms the widely held view that leading the league in *RBI*s gives a player a big advantage toward getting the Most Valuable Player (MVP) award. Leading in *AVG*, *SLG*, or other categories are less important, unless of course a player leads in several of these categories. By far the second most important accomplishment for the MVP is playing for a league or division championship team. See also, "Biases in MVP Voting," in [26, *Win Shares*], page 194.

References

Jim Albert, *Comment on: A statistical analysis of hitting streaks*, J. American Statistical Association, (1993), 1184–1188.

Jim Albert, *The home-run hitting of Mike Schmidt*, Chance, (1998), 27–34.

Jim Albert and Patricia Williamson, *Using model/data simulations to detect streakiness*, American Statistician, (2001), 41–50.

Christian Albright, *A statistical analysis of hitting streaks*, J. American Statistical Association, (1993), 1175–1183.

Scott M. Berry, *Does "the zone" exist for homerun hitters?*, Chance, 12 (1999), 51–56.

J. Eric Bickel and Dean Stotz, *Batting average by count and pitch type: Fact and fallacy*, The Baseball Research Journal, **31** (2003), 29–34.

Phil Birnbaum, *Does a pitcher's stuff vary from game to game?*, The Best of By the Numbers, (2003), 101–105. Originally published 2000.

Phil Birnbaum, *The run value of a ball and strike,* The Best of By the Numbers, (2003), 28–30. Originally published in 2000.

Clifford Blau, *Do faster runners induce more fielding errors?,* The Best of By the Numbers, (2003), page 41. Originally published in 1999.

Clifford Blau, *Offensive replacement levels,* The Best of By the Numbers, (2003), 53–55. Originally published in 1998.

Harold Brooks, *Playing every day and September performance,* The Best of By the Numbers, (2003), 80–83. Originally published in 1992.

Clem Comly, *Average Run Equivalent Method,* The Best of By the Numbers, (2003), 32–37. Originally published in 2000.

Andrew Cornish, Charles Fiedler, Brian S. Foss, and Paul M. Sommers, *Eye-pothesis testing: Another look at the 1986 World Series,* Journal of Recreational Mathematics, **19** (1987), 241–246.

Reid Dorsey-Palmateer and Gary Smith, *Bowlers' hot hands,* American Statistician, 58 (2004), 38–45.

Michael Freiman, *56-Game hitting streaks revisited,* SABR Baseball Research Journal, (2003), 11–15.

Tom Hanrahan, *A new way of platooning: Ground ball/fly ball,* The Best of By the Numbers, (2003), 95–99. Originally published in 1999.

Tom Hanrahan, *Catchers: Better as veterans,* The Best of By the Numbers, (2003), 12–19. Originally published in 1999.

Tom Hanrahan, *Does good hitting beat good pitching?,* The Best of By the Numbers, (2003), 46–51. Originally published in 2001.

Keith Karcher, *Winning streaks, losing streaks, and predicting future team performance,* The Best of By the Numbers, (2003), 91–93. Originally published in 1989.

Dan Levitt, *Do faster runners induce more fielding errors?,* The Best of By the Numbers, (2003), 39–40. Originally published in 1998.

Sig Mejdal, *The recipe for a stolen base,* The Best of By the Numbers, (2003), 76–78. Originally published in 2000.

Pete Palmer, *Clutch hitting one more time,* The Best of By the Numbers, (2003), 43–44. Originally published in 1990.

Mark D. Pankin, *Finder better batting orders,* The Best of By the Numbers, (2003), 57–65. Originally published in 1991.

Tom Ruane, *Stolen base strategies revisited,* The Best of By the Numbers, (2003), 21–26. Originally published in 1999.

Paul M. Sommers, *Home-field advantage in the World Series: Myth or reality,* Journal of Recreational Mathematics, **28** (1996–1997), 180–184.

Paul M. Sommers, *Sultans of swat and the home runs test,* Journal of Recreational Mathematics, **30** (1999–2000), 118–120.

Rob Wood, *How often does the best team win the pennant?,* The Best of By the Numbers, (2003), 85–89. Originally published in 2000.

Rob Wood, *What drives MVP voting?,* The Best of By the Numbers, (2003), 67–74. Originally published in 1999.

The 32nd issue (2004) of SABR's *The Baseball Research Journal* just arrived. Among many interesting articles is one by Herm Krabbenhoft concerning hitter's streaks of consecutive games getting On Base. Joe DiMaggio's 1941 streak of 56 consecutive games with a hit extends to 74 consecutive games getting on base. An extensive search shows that this is the second longest such streak, the longest being 84 games by Ted Williams in 1949.

Appendix
2

The Binomial Theorem

If you followed the calculations in this book involving the numbers $_nC_k$, and if you would enjoy seeing a connection to algebra, this appendix is for you.

If you multiply out $(x + y)^2$, $(x + y)^3$, $(x + y)^4$, and $(x + y)^5$ and collect terms, you will get

$$(x + y)^2 = x^2 + 2xy + y^2,$$
$$(x + y)^3 = x^3 + 3x^2y + 3xy^2 + y^3,$$
$$(x + y)^4 = x^4 + 4x^3y + 6x^2y^2 + 4xy^3 + y^4,$$

and

$$(x + y)^5 = x^5 + 5x^4y + 10x^3y^2 + 10x^2y^3 + 5xy^4 + y^5.$$

Note that the end terms, like x^4 and y^4, have understood coefficients 1 and that the next-to-end terms, like $4x^3y$ and $4xy^3$, have coefficient equal to the power of $(x + y)$ being calculated. The other coefficients may be more mysterious. All of these expansions are special cases of the Binomial Theorem, which follows.

Binomial Theorem

If x and y are any numbers, and if n is a whole number, then $(x + y)^n$ is equal to

$$(x + y)^n = x^n + nx^{n-1}y + {}_nC_{n-2}x^{n-2}y^2 + \cdots$$
$$+ {}_nC_2x^2y^{n-2} + nxy^{n-1} + y^n.$$

Since ${}_nC_n = 1 = {}_nC_0$ and $y^0 = x^0 = 1$, the end terms are equal to ${}_nC_n\, x^n\, y^0$ and ${}_nC_0 x^0 y^n$. Since ${}_nC_{n-1} = n = {}_nC_1$, the next-to-the-end terms are equal to ${}_nC_{n-1}x^{n-1}y^1$ and ${}_nC_1x^1y^{n-1}$. In other words, $(x + y)^n$ is the sum of all terms ${}_nC_kx^ky^{n-k}$, where k runs over the values $0, 1, 2, \ldots, n$.

The expression $x + y$ is called a "binomial." This is where the theorem gets its name. As noted on page 43, the numbers ${}_nC_k$ are often called "binomial coefficients" and written $\binom{n}{k}$.

Note that ${}_4C_2 = \frac{4 \times 3}{2 \times 1} = 6$; this explains the coefficient of x^2y^2 in the expansion of $(x + y)^4$. The coefficients of x^3y^2 and x^2y^3 in $(x + y)^5$ are ${}_5C_2 = \frac{5 \times 4}{2 \times 1} = 10$ and ${}_5C_3 = \frac{5 \times 4 \times 3}{3 \times 2 \times 1} = 10$. This symmetry always holds:

$$_nC_{n-k} \quad \text{always equals} \quad {}_nC_k.$$

Here are two interesting special cases that we've more-or-less already encountered in this book. If we set $x = p$ and $y = q$ where $p + q = 1$, then the Binomial Theorem gives

$$1 = (p + q)^n = p^n + np^{n-1}q + {}_nC_{n-2}p^{n-2}q^2 + \cdots$$
$$+ {}_nC_2p^2q^{n-2} + npq^{n-1} + q^n.$$

As noted in Chapter 6, for each k the number ${}_nC_kp^kq^{n-k}$ is the probability of exactly k successes in n Bernoulli trials.

The sum runs over the values $k = 0, 1, 2, \ldots, n$. Because precisely one of the events "exactly k successes" must occur, the sum of these probabilities must be 1, by Property 2 on page 22. The Binomial Theorem gives an algebraic explanation of the same equation.

Now set $x = y = 1$ in the Binomial Theorem. You will get

$$2^n = 1 + n + {}_nC_{n-2} + \cdots + {}_nC_2 + n + 1. \qquad (*)$$

This time, the left-hand side can be viewed as the number of strings of successes and failures of length n. This is because there are two choices for the first outcome, S or F, two choices for the second outcome, and so forth. So there are

$$2 \times 2 \times 2 \times \cdots \times 2 \quad [n \text{ times}] = 2^n$$

such strings of length n. A typical term on the right-hand side of the equation $(*)$ is ${}_nC_k$, which is the number of such strings with *exactly* k Ss, and hence exactly $n - k$ Fs. Again, the sum runs over $k = 0, 1, 2, \ldots, n$. So the equation $(*)$ says that the number of strings of S and F of length n is just the sum of the number of all such strings with exactly k Ss, as k runs over $k = 0, 1, 2, \ldots, n$.

The probabilities for the lengths of World Series given in the table on page 95 need to add to 1:

$$p^4 + q^4 + 4(p^4q + pq^4) + 10(p^4q^2 + p^2q^4) \\ + 20p^3q^3 = 1. \qquad (**)$$

Here is how to verify this algebraic equation. Since $p + q = 1$, $p^4 + q^4$ can be replaced by $(p^4 + q^4)(p + q)^2$ and $4(p^4q + pq^4)$ can be replaced by the product

$4(p^4q + pq^4) (p + q)$. If you make these replacements in the left-side of equation (**), and then carefully multiply out and collect like terms, you will get

$$p^6 + 6p^5q + 15p^4q^2 + 20p^3q^3 + 15p^2q^4 + 6pq^5 + q^6.$$

Now by the Binomial Theorem, this is exactly $(p + q)^6$. And since $p + q = 1$, this is equal to 1, so equation (**) is correct. It is *not* a general identity, but it does hold if $p + q = 1$.

Here is why the Binomial Theorem is true. Consider the operation of multiplying out

$$(x + y)^n = (x + y)(x + y)(x + y) \times$$
$$(x + y)(x + y) \cdots (x + y).$$

The product on the right is the sum of all products obtained by choosing x or y from the first factor, choosing x or y from the second factor, choosing x or y from the third factor, etc. The number of appearances of the term $x^k y^{n-k}$ is just the number of ways of choosing x exactly k times, i.e., $_nC_k$. So, when you collect terms, the coefficient of $x^k y^{n-k}$ will be $_nC_k$.

Annotated Bibliography

The pioneering book that indicated that baseball would be more efficient if managers used statistics instead of just hunches was published in 1964. This is the book [10, *Percentage Baseball*] by Earnshaw Cook. It is interesting historically, but it is an annoying and not very useful book. And it's not easy reading. In my personal notes, I wrote "I find the author pompous and sarcastic and substantially insulting to baseball people. There's a lot of repetition in his mathematics and his snide remarks." When I wrote that, I thought I was being harsh. So it was heart-warming to read the following in [30, *Moneyball,* page 71]: "A professor of mechanical engineering at Johns Hopkins, Earnshaw Cook, wrote two pompous books, in prose crafted to alienate converts, that argued for the relevance of statistical analysis in baseball." Nicely put. Beyond his style, my main concern with Cook's book is that the author believes too much in the statistics. The real baseball world is messier, which is what makes it so fascinating. It is interesting to compare what the idealized statistics predict with what really happens. The author takes the data of the past and then says that it ought to dictate the future. He especially thinks baseball people are ignorant fools for using their hunches instead of the "facts." I'm not denying that there's some truth to the author's concerns. More knowledge of probability would undoubtedly help improve the managers' hunches, but I have a lot more respect for hunches and intuition than the author does.

[1] Albert, Jim. *Teaching Statistics Using Baseball,* Mathematical Association of America, 2003. Ideal text for a freshman-level statistics course based on baseball.

[2] Albert, Jim and Jay Bennett. *Curve Ball,* 2nd edition, Copernicus Books, Springer-Verlag New York, Inc., 2003. A fine book that would be suitable for supplementary reading in a course using [1].

[3] Angell, Roger. *The Summer Game*, Viking Press, New York, 1962. A wonderful book that gives a lot of inside information even though it was published in 1962.

[4] Armour, Mark L. and Daniel R. Levitt. *Paths to Glory,* Brassey's, Inc., Washington, DC, 2003. An excellent book: well-written, thoroughly researched and fascinating. Like [36], the focus is on dynasties and would-be dynasties, but the emphasis is on how these teams "got that way" or how they fell short of expectations. The appendices would be especially valuable to readers of my book or users of the "bibles" [23], [24], [25], and [43].

[5] Barra, Allen. *Clearing the Bases,* Thomas Dunne Books, St. Martin's Press, New York, 2002. Concerns "the Greatest Baseball Debates of the Last Century" and has some interesting stories.

[6] Bernstein, Peter L. *Against the Gods,* John Wiley & Sons, Inc., New York, 1996. This book is more concerned with the gambling and risks in the financial world, but covers many many topics.

[7] Berra, Yogi. *The Yogi Book: "I really didn't say everything I said."* Workman Publishing Co., New York, 1998.

[8] Bouten, Jim. *Ball Four,* The World Publishing Co., New York, 1970. Up until 1970, sportswriters and authors treated players with great reverence, and they avoided publicizing their warts, i.e., their questionable behavior off the job. In those days, a similar restraint applied to politicians. All this changed, in the setting of baseball, with the publication of this book. This very funny book is candid and revealing. Heroes are shown to be just as subject to flaws as the rest of us. Readers loved this best-seller, but many baseball people were appalled. If you haven't read this classic, start by reading the preface to Bouten's follow-up book [9].

[9] Bouten, Jim. *I'm Glad You Didn't Take It Personally,* A Dell Book, New York, 1971. This is dedicated to two of his most vociferous critics, Dick Young and Bowie Kuhn.

[10] Cook, Earnshaw. *Percentage Baseball,* MIT Press, Boston, 1966.

[11] Costas, Bob. *Fair Ball,* Broadway Books, New York, 2000.

[12] Dawidoff, Nicholas. *The Catcher Was a Spy,* Vintage Books, Random House, Inc., 1994. About a fascinating guy, Moe Berg, who was a mediocre Major League ballplayer for 15 years prior to World War II and then became a controversial spy during the war.

[13] De Vito, Carlo. *The Ultimate Dictionary of Sports Quotations,* Facts on File, Inc. New York, 2001.

[14] Dickson, Paul. *Baseball's Greatest Quotations,* Harper Perennial, New York, 1991. A second edition is in the works.

[15] Edwards, Bob. *Fridays with Red,* Simon & Schuster, 1993. The title refers to Bob Edwards' friendship with Red Barber during 12 years they did a radio program together. But the book is a lot more; it's a tribute to Red's life and includes a lot of his great quotes.

[16] Feller, William. *An Introduction to Probability Theory and Its Applications,* volume 1, 3rd edition, John Wiley &

Sons, Inc., New York, 1967. Ideal for sophisticated readers who are able to study science or mathematics on their own. The level is a roller coaster, with difficult sections cheerfully intermixed with easier sections without warning. The first edition, published in 1950, had a big impact on the theory of probability in the United States because it was the first accessible, but serious, probability book in English.

[17] Goodwin, Doris Kearns. *Wait Till Next Year: A Memoir,* Simon & Schuster, New York, 1997. A sweet story about a little girl growing up near New York City as a baseball fan during almost the same era that Kahn's book covers: 1949–1957. Those were trying times to be a Dodger fan.

[18] Griffin, Peter. *Extra Stuff: Gambling Ramblings,* Huntington Press, Las Vegas, 1991.

[19] Griffin, Peter. *The Theory of Blackjack,* 5th edition, Huntington Press, Las Vegas, 1996. The author's specialty is blackjack. He has analyzed the probabilities and strategies of blackjack very thoroughly, both in this book and in [18].

[20] Halberstam, David. *Summer of '49,* William Morrow and Co., Inc., 1989. A great story about the 1949 pennant race between the traditional long-time rivals, the Boston Red Sox and the New York Yankees, which wasn't decided until the final game of the season.

[21] Halberstam, David. *The Teammates,* Hyperion Books, The Amateurs, Inc., New York, 2003. About the last days of Ted Williams when he was visited by his old baseball buddies. It's not the greatest literature, but it is a warm and fuzzy story.

[22] Hoban, Michael. *Baseball's Complete Players,* McFarland & Co., Inc., Jefferson, NC, 2000. This book rates non-pitcher players using his Hoban Effectiveness Quotient (HEQ). The book is too personalized, and the definition of HEQ seems arbitrary. Incidentally, HEQ is the sum of products, not a quotient.

[23] James, Bill. *Major League Handbook,* STATS, Inc., Morton Grove, IL. This was published annually, but has been superseded by *The Bill James Handbook,* ACTA Publications, 2003.

[24] James, Bill. *All-Time Baseball Sourcebook,* STATS, Inc., Skokie, IL, 1998.

[25] James, Bill. *The New Bill James Historical Baseball Abstract,* Free Press, New York, 2001.

[26] James, Bill and Jim Henzler. *Win Shares,* STATS, Inc., Skokie, IL, 2002.

[27] Kahn, Roger. *The Era 1947–1957,* University of Nebraska Press, Lincoln, 1993. Covers the decade when the three teams in New York (the Yankees, the Giants, and the Dodgers) "ruled the World." This book, published in 1993, is well-written and insightful, and the author doesn't pull punches.

[28] Lang, Arne K. *Sports Betting 101,* Gamblers Book Club/ GBC Press Inc., Las Vegas, 1992. The subtitle of this book is "Making Sense of the Bookie Business and the Business of Beating the Bookie."

[29] Levinson, Horace C. *Chance, Luck and Statistics,* Dover Publications, Inc., Mineola, NY, 1963. Originally published in 1939.

[30] Lewis, Michael. *Moneyball,* W. W. Norton & Co., New York, 2003. A well-written account of how the Oakland A's management, and in particular the manager Billy Beane, uses statistical analyses (especially the on base percentage) to find promising ballplayers that other franchises have shown little interest in. Another strategy is to focus on prospects in college rather than right out of high school. The proof of the pudding is that for several years the A's have fielded remarkably strong teams with relatively low payrolls. Some sportswriters and people in baseball have reacted very strongly against this book, but I think it is

good and fair, though he does tend to over-idolize Billy Beane and his computer-geek assistants.

[31] Liebman, Glenn. *1001 Baseball Quips and Quotes,* Gramercy Books, Chicago, 1994.

[32] Liebman, Glenn. *Grand Slams!,* Contemporary Books, Chicago, 2001.

[33] Moore, David S. *The Basic Practice of Statistics,* W. H. Freeman and Co., 1995. My favorite freshman-level statistics text.

[34] Morris, Jim and Joel Engel. *The Oldest Rookie,* Little, Brown and Company, Boston, 2001. About Jim Morris's amazing story leading up to a Major League career that began at age 35.

[35] Nathan, David H. *Baseball Quotations,* Ballantine Books, New York, 1991.

[36] Neyer, Rob and Eddie Epstein. *Baseball Dynasties,* W. W. Norton & Co., New York, 2000. Analyzes 15 of the greatest dynasties in baseball. They use the usual familiar criteria, including the number of pennants won and the number of World Series won. But they rely heavily on what they call an "SD score," which is an unbiased way to measure how much better or worse a team is than others of the *same* year or span of years. They also promise to give their opinion as to the greatest team of the 20th century. I had to wait until the end of the book to find out; so I'm not telling you.

[37] Oh, Sadahiro and David Falkner. *Sadahiro Oh: A Zen Way of Baseball,* Times Books, New York, 1984. A fascinating book about a great player from another culture.

[38] Reifman, Alan. *Hot Hand in Sports,* http: //www.hs.ttu .edu/hdfs3390/hothand.htm. A huge and very active Website, with links to all sorts of articles dealing with streaks in various sports.

[39] Ross, Sheldon. *A First Course in Probability,* 4th edition, Macmillan College Publishing Co., New York, 1994. An

excellent source for a serious introduction to probability. This book is used as a text for junior and senior level college courses in probability theory.

[40] Runquist, Willie. *Baseball by the Numbers,* McFarland & Co., Inc., 1995.

[41] Schell, Michael J. *Baseball's All-Time Best Hitters,* Princeton University Press, Princeton, NJ, 1999. This book is difficult and sophisticated. The author adjusts Major League batting averages to take into account things like the era, the competition, and the ballparks. It's a scholarly study, but focuses entirely on batting average. As we know, there are several other measures that more accurately provide the value or overall effectiveness of ballplayers.

[42] Thorn, John and Pete Palmer, *The Hidden Game of Baseball,* Doubleday, New York, 1985.

[43] Thorn, John and Pete Palmer, editors. *Total Baseball,* 2nd edition, Warner Books, Inc., New York, 1991.

Acknowledgments

In addition to the standard general references for quotations, such as Bartlett's Famous Quotations, I used the following sources: [7, *The Yogi Book*], [13, *The Ultimate Dictionary of Sports Quotations*], [14, *Baseball's Greatest Quotations*], [31, *1001 Baseball Quips and Quotes*], [32, *Grand Slams!*], and [35, *Baseball Quotations*].

First, I'm grateful to everyone in my extended family. My wife, Ruth, has been a supporter of all of my projects, and this has been no exception. She's been a wonderful sounding board and grammar consultant. My daughters, Laurel and Emily, reminded me of how they viewed baseball games when they were little, and Emily provided some of the prose in the Preface. She also helped me sort out the quotations. Her husband, Ted Clifford, provided general advice and enthusiasm.

My stepdaughter, Lisa Madsen, gave a critical reading of an early manuscript and helped me change my voice from that of a professor to that of a baseball fan. Her husband, Dan Dalthorp, offered good advice and also helped me find the Simpson's Paradox examples. Finally, I'd like to thank my granddog, Mary, for not voicing an opinion about the title of the book.

I thank Don Albers and Ann Daniel for leading me to my excellent editor, Stephen Morrow. Stephen is a conscientious slave-driver. He kept telling me he liked my voice, but he gave me voice I didn't even know I had. It's been a pleasure to work with him.

This book is an unrecognizable outgrowth of a freshman seminar, "Statistics and Mathematics of Baseball," that I taught in 2000. I appreciate the students who tolerated the statistics as well as the fact that I didn't know enough baseball. I thank all the guys in our local Baseball Book Club who have been supportive, especially Mark Armour and Dan Schlewitz.

Other folks who made helpful contributions were Jim Albert, Art Benjamin, Steve Bleiler, Barry Garelick, Jeanie Neven, Martha Osgood, Doug Pappas, and Hao Wang.

Index